CW01463786

A HIMALAYAN ORNITHOLOGIST

BRIAN HOUGHTON HODGSON

A HIMALAYAN ORNITHOLOGIST

The life and work of
BRIAN HOUGHTON HODGSON

MARK COCKER

AND

CAROL INSKIPP

COLOUR PLATES OF PAINTINGS FROM
THE COLLECTION OWNED BY
THE ZOOLOGICAL SOCIETY OF LONDON

Oxford   New York   Tokyo
OXFORD UNIVERSITY PRESS
1988

Oxford University Press, Walton Street, Oxford OX2 6DP
Oxford New York Toronto
Delhi Bombay Calcutta Madras Karachi
Petaling Jaya Singapore Hong Kong Tokyo
Nairobi Dar es Salaam Cape Town
Melbourne Auckland
and associated companies in
Berlin Ibadan

Oxford is a trade mark of Oxford University Press

Published in the United States
by Oxford University Press, New York

© (text) Mark Cocker and Carol Inskipp, 1988
© (plates) Zoological Society of London, 1988

All rights reserved. No part of this publication may be reproduced,
stored in a retrieval system, or transmitted, in any form or by any means,
electronic, mechanical, photocopying, recording, or otherwise, without
the prior permission of Oxford University Press

British Library Cataloguing in Publication Data
Cocker, P. Mark
A Himalayan ornithologist: the life and
work of Brian Houghton Hodgson.
1. Asia. Himalayas. Exploration.
Hodgson, B. H. (Brian Houghton), 1800–1894
I. Title   II. Inskipp, Carol
915.4'0431'0924
ISBN 0–19–857619–6

Library of Congress Cataloging in Publication Data
Data available

Set by Rowland Phototypesetting Ltd
Bury St Edmunds, Suffolk

Printed in Great Britain
at the University Printing House, Oxford
by David Stanford
Printer to the University

THIS BOOK IS DEDICATED TO
OUR RESPECTIVE PARENTS
TOM AND JOYCE ROBINSON
PETER AND ANNE COCKER

ACKNOWLEDGEMENTS

The author Alan Octavian Hume, who dedicated his book to Hodgson, said that 'a work of this kind requires leisure; time to consult all that has ever been written by others . . . time to weld all this together with personal and unpublished experience into a harmonious whole; time to rewrite and revise and polish.'

Like Hume, the authors of this Hodgson book have often felt the need for time and leisure during the course of its preparation. Fortunately, whenever these commodities have seemed in short supply our friends, relatives, and colleagues have invariably come to our assistance. We would therefore like to take this opportunity to thank them for their time and effort on our behalf, without which the whole project would have been much more difficult.

In particular, we are grateful to Joyce Robinson and Sue O'Connell for struggling with our atrocious handwriting, and turning chaos into order; to Tim Inskipp for his invaluable efforts in obtaining many references; and to him, Mary Muir, and Joyce and Tom Robinson for their constant help and support. We wish to thank Richard Grimmett for helping to set the project afloat, and then Gerald Crowson, John Inskipp, Paul Lewis, Mary Muir, and Joyce Robinson for their helpful comments upon earlier drafts, to Dr Tim Sharrock for his encouragement in the early stages; and finally thanks go to Tim Inskipp and Joyce and Tom Robinson for their painstaking checking of the final typescript. We acknowledge the valuable advice given to us by John Denson and Mark Holmstrom; then the staffs of the Zoological Society of London, especially the Librarian, Reg Fish, The India Office Library, and The Zoology Library of the British Museum (Natural History) for their generous assistance.

TAXONOMIC NOTE

Both the taxonomic order of families and the English and scientific names of birds follow Inskipp and Inskipp (1985).[129] The English and scientific names of mammals follow Corbet and Hill (1980).[10]

CONTENTS

PLATES

MAMMALS

I

The longest-lived of them all

Exactly one month after the dawning of the nineteenth century, Brian Houghton Hodgson was born to a middle-class family in the parish of Prestbury, Cheshire. When he was still a young boy his father, a banker in Macclesfield, had lost the family fortunes in Irish mining ventures. The house where he was born on 1 February 1800 was closed down and the family moved to Macclesfield, then Congleton, and finally to Essex, where Brian's father took up a military position from 1814 to 1820. In spite of these straitened times, the first son of an eventual family of seven continued his education uninterrupted, firstly in Macclesfield then later in Richmond, Surrey.

Brian was an intelligent boy, and through a family friend, James Pattison, then one of the directors of the East India Company, he secured nomination to the Company's civil service, and entered Haileybury College in 1816. This institution had been established to train the growing number of administrators of a commercial company that was fast becoming the dominant military and political power in the Indian subcontinent. It had attracted some of the period's leading intellectuals, including Thomas Malthus, the brilliant professor of political economy, and the Reverend Henry Walter, previously a professor of chemistry at Cambridge. It was probably the latter who instilled in the young Hodgson a love of natural history, while the celebrated economist, then at the height of his fame, cultivated in his ward deep liberal convictions in politics. Hodgson was allowed to lodge with Malthus, whose house had become a resort for the leading Whigs of the day. In the college he soon demonstrated a capacity for scholarship, carrying off the first term's prize for Bengali, and then in the second term, the Classics prize. In December 1817, Hodgson left Haileybury (gold medallist for his year) and awaited his call from India.

In the early nineteenth century all the East India Company's trainee officials were given a year's tuition in the native languages and the law of India, at Calcutta's College of Fort William. Hodgson again completed his studies with distinction, and through the influence of family friends, Sir Charles and Lady D'Oyly, he was introduced to the leading dignitaries of

Calcutta. However, in 1819 Hodgson fell victim to the fever that was to decide so much about his later life and at the age of 19 presented him with little choice—'a grave on the plains, or a post in the hills'.[127]

With the help of his powerful connections Hodgson soon secured the latter under the Commissioner of Kumaon, George William Traill, a strong administrator often known as the 'King of Kumaon'. Traill was just the man to help consolidate the Company's position after Lord Hastings, the Governor-General, had successfully campaigned against three strong, native powers—the Marathas, the Pindaris, and the Nepalese. Two years of hill fighting against the last had left Major General Ochterlony the clear victor, and enabled him to force upon the Nepalese, the Peace of Segouli in 1816. Kumaon, on the western border of Nepal, had previously been absorbed by its neighbours, but was ceded back to the British along with other territory. It was now Traill's job to enforce British rule and to place on a clear and firm footing administration and tax collection.

Working with this young and vigorous superior, Hodgson was given an excellent training in the administration of a distant territory. His health responded to the more genial hill climate and, during the course of his duties, which involved a full assessment of the district's taxable revenue, he became an expert mountaineer. Traill and Hodgson's work in this field was written up into a report in 1819 that formed the basis of a handbook on the region over fifty years later, so thoroughly was it completed. Traill inculcated in his assistant a belief that a frontier administrator, in the interest of a land which only he as its direct governor could fully understand, might sometimes have to disregard the orders of a distant central authority. This was a principle which informed much of Hodgson's later diplomatic career at the court of Kathmandu in Nepal.

In 1820, it was to this country that Hodgson eventually went as assistant resident under another able administrator, Edward Gardner. After the Peace of Segouli it had been Gardner's task to improve relations between the British and the Nepalese, largely by a policy of careful non-intervention in that country's domestic affairs. The passive role on the part of the British in Kathmandu contrasted sharply with Hodgson's dynamic life in the hills of Kumaon. He soon found himself with little to do, and frustrated by inactivity he welcomed the opportunity offered by a post in Calcutta.

When Hodgson took up his position as deputy secretary in the Persian Department, the most coveted of junior political positions, it was obvious to those around him that he was destined for a distinguished political career. In five years of service he had already given abundant evidence of his talents; surely the new post would be a stepping stone to a position of real importance in the British Indian administration? Yet fate can often decree that weakness rather than merit determines the course of our lives. Hodgson's weakness was a susceptibility to the climate of Bengal: he had fallen seriously ill once again. It is difficult to diagnose exactly his 'oriental fevers', but he probably suffered from a combination of the intestinal and hepatic disorders that afflict contemporary Europeans in the Indian subcontinent, and as now, the cooler and less humid climate of the Himalayan foothills favoured recovery.

The assistant residentship was occupied when Hodgson returned to Kathmandu, so he was put in charge of the post office there until the existing assistant left. That happened in 1825: Hodgson was back, this time to stay.

Baulked in his professional ambitions, and exiled as Hodgson was, a lesser man might have slipped into a weary cynicism. A land of tough hill tribes, proud of their warrior spirit, suspicious of British intrusion, there seemed little in Nepal to attract a civilized gentleman like Hodgson. An ancient Chinese tradition, to whose government the Nepalese durbar paid token obeisance, forbade the presence of a white woman, since it was prophesied that this would bring downfall to the empire. So the small British community in Kathmandu was without its better half, and by edict of the government, confined to the Kathmandu Valley, only 22 km across at its widest. The non-interventionist policy of the British government left its officials there with few duties, and most enjoyed a condition of enforced idleness. Of course, there were always sports and hunting, the great colonial pursuits, to relieve the boredom, and Hodgson was keen on both, particularly shooting for pheasant and woodcock. Yet Hodgson was no waster of time. Set free in this virgin terrain, he was determined to study it, taking off in all directions.

As early as 1824 he had begun to employ local hunters to collect specimens of the area's birds and mammals, while training local artists to sketch and paint them. In spite of his assistants, Hodgson himself liked to make careful analyses of the internal anatomy of the various animals that were brought to him. He also supervised personally their illustration and took extensive notes on his discoveries, while over a dozen scholarly assistants helped in the preparation and translation of his growing collection of Sanskrit manuscripts. By 1826 he had written his first paper on the origins and affinities of the various Himalayan tribes, but he was also interested in Himalayan languages, Buddhism, geography, architecture—in fact, anything new that fell before his eye. His collection of Sanskrit manuscripts eventually developed into a comprehensive canon of Buddhist writings and prompted him to undertake the first European analysis of Northern or Mahayana Buddhism. Hodgson also made extensive enquiries into Nepalese military affairs, its legal system and its trade.

Then, of course, there were his professional duties, which increased greatly in 1829 when Edward Gardner gave up his post in Nepal and left Hodgson to officiate in his absence. While he was still only in his twenties the Court of Directors was unwilling to entrust the resident's full title to one so young. Yet he was being groomed for the position, and when Herbert Maddock took over from Gardner in Kathmandu after a two-year gap he wrote in high praise of Hodgson's talents. It was obvious to Maddock that he was fully qualified to become the Company's representative in Nepal, and when Maddock left in 1833 to take up the important residentship of Oudh, an Indian state on Nepal's southern border, Hodgson was his successor in Kathmandu.

With his new position he commanded an annual income of £4000, a substantial figure in early nineteenth-century India. Yet demands were made upon his purse from all quarters. There were his private researches to finance, and his pension of £1000 after 25 years service to maintain. His family, too, needed considerable financial support. Both his brothers were educated at Haileybury College: the elder, William, five years Brian's junior, joined the Bengal Artillery Regiment; while Edward, the youngest of the seven siblings, began his professional life in the Company's civil service. In both their education and their respective careers the eldest brother in Nepal would have to provide for their promotion; Hodgson did so with loyalty and generosity. In 1834 when William fell ill, Brian looked after the invalid in his own house for almost a year.

It was not the expense that troubled Hodgson, but he did complain to his family about the increasing formality of their letters and their almost reverent tone for a son and brother whom they could hardly remember, but whose support was the mainstay of their lives. Severed from home life since the age of 18, and denied the social intercourse which a larger British community might have provided, Hodgson felt starved of love and affection. It particularly piqued him that his sister Frances—'darling Fanny'—should be free and affectionate in her letters to William, but formal and awkward to him! A letter from James Prinsep, the secretary of the Bengal Asiatic Society, written by his female secretary, was enough to awaken in Hodgson his want of the opposite sex. His reply eulogises: 'I have your letter in some pretty lady's hand! Oh but it's a rare thing to have those sweet helpmates about one!' Even the gentleman scholar felt pinched by Nepal's embargo on white women.[36]

In 1837 Hodgson entered into a relationship with a Kashmiri Moslem lady, an affair about which he was quite open, an uncharacteristic attitude amongst the Indian Britons of the period. Their union leavened his physical and spiritual isolation and gave him two children of his own who were brought up in Holland under the direction of Brian's eldest sister.

Yet misfortune had not done with the Hodgson family. The fever which had played such a large part in shaping the course of Brian's own life, in July 1835 ended that of his brother Edward. A snipe-shoot in the swamps during monsoon had brought on a severe condition, and the young man, who had shown great promise and was the Assistant Commissioner of Meerut when still only 22, died soon after. In 1837 Hodgson himself suffered a recurrent bout of his old sickness and had set out on a visit to a specialist in Calcutta when the tumour in his liver burst and he recovered. By March 1838 he had returned to Nepal, and had begun to find his old form, only to have his other brother, William, die three months later on 12 June 1838. In both cases Brian was obliged to cover their debts, a business which he had always done uncomplainingly.

From this early period of Hodgson's career in Nepal the picture that emerges is of a serious, lonely, and rather sad figure. His prime was spent almost entirely in the foothills of the Himalayas, away from the society of his family and friends, and the letters he sent betray a yearning for sympathy and affection, particularly the softening touch of female company. As a result of the political isolation of the British in Kathmandu his life was largely one of petty domestic affairs, and the fragility of his health imposed upon that a strict, almost ascetic regime. Already predisposed to pride, Hodgson's solitude deepened this quality into stiffness and hauteur. However, he was a generous soul, giving freely from his private purse, and from his huge collections of manuscripts and natural historical specimens. Finally, and most importantly, his supreme knowledge of Nepalese affairs and culture, sifted during these twelve quiet years of patient enquiry, rendered him the best man to weather a succession of storms that burst over Nepal during his later years in office.

A RESIDENT IN NEPAL

Running along the line of the greatest mountain system in the world, Nepal is a small Himalayan kingdom, roughly rectangular in shape some 835 km long and 340 km at its widest. Squeezed on two sides by giant empires—India to the south and China to the north—it might seem remarkable that during the eighteenth century Nepal experienced a period of rapid expansion under the newly established Gurkha dynasty. Its leader, Prithi Narayan Shah, traditionally of Rajput descent, was a warrior of some skill who succeeded in welding together the numerous small city-states and hill strongholds into a unified country much the same size as it is today. His people, a racial admixture of Aryan Rajputs from the south and Mongol hill tribes, were fiercely independent and deeply proud of their martial traditions. Checked in their territorial ambitions to the north by a decisive Chinese defeat in 1792, they focused their attention with greater success on the hill districts at their flanks. In the west they first absorbed the territory of Kumaon, and in the east encroached upon the independent kingdom of Sikkim. However, it was to the south and the prosperous Gangetic plain with its rich hinterland that Nepalese leaders looked with the keenest eye.

Suspicious of Company strength, they nibbled only piecemeal at the lowlands immediately south of the Himalayan foothills, in the hope that they might avoid direct confrontation. This narrow belt of land, no greater than 32 km at its widest, and known as the terai, was an area of dense jungle, notorious for its deadly malaria or 'awal'. It was, however, extremely fertile and much coveted by the Nepalese as a source of agricultural revenue and as *Lebensraum* for its mountain-locked population. Their insignificant but constant expansion enabled the Nepalese to take much valuable land but forced matters to a head between themselves and 'John Company' in 1814.[140]

Bhim Sen Thapa, the strong and skilful chief minister of Nepal since 1804, had miscalculated the British response to his gradual infiltration of the terai, but he was in no doubt about the strength and capacity of the East India Company forces. He had carefully watched the British system of protectorates and the sequence of events that attended them—first commercial, then military presence, and finally British conquest: he was determined that this should not happen to Nepal. An earlier British mission to Kathmandu headed by Captain Knox in 1802, had made little progress at the Nepalese durbar and retired to Company territory the following year. It was only with the Peace of Segouli in 1816, and the conclusion of an Anglo-Nepalese war which Bhim Sen had tried repeatedly to settle by diplomacy, that a resident was successfully imposed upon the court of Kathmandu. However, the chief minister was determined to isolate the British contingent in his capital, and to frustrate Company involvement in Nepalese domestic affairs. It was this state of political impasse that Brian Hodgson found when he arrived there in 1820 as Edward Gardner's assistant.

Although the war and the British annexation of the terai in 1816 had seriously deflated the aspirations of the war party in Kathmandu it had not completely disarmed it. Bhim Sen Thapa with characteristic wisdom, realized that in order to ensure the security of the Nepalese state he had to play a double game. On the one hand he must provide some natural outlet for the bellicose spirit of his fellow countrymen, while, on the other, refusing to give offence to the British. Further conflict was to be avoided at all costs, for the last war had demonstrated to him, if to no one else, that in open battle

Himalayan ardour was no match for the discipline of the British regiments. With the staunch support of the Queen Regent, Maharani Lalit Tripuri Devi he could steer a firm course in the Nepalese durbar and at the same time keep the British resident at arm's length from both the royal court and the government. The resident, Edward Gardner, for his part was willing to acquiesce in this situation in order to establish more amicable links between the two powers. He was also prepared to turn a blind eye to Nepal's constant military operations as long as they were confined to the parade ground and ceremonial display. While this tacit agreement between individuals existed there could be an uneasy concord between the two powers.

During his early years in Nepal, Hodgson recognized the unstable nature of this personal government, and by a systematic inquiry into the military, judicial, and commercial affairs of Nepal he hoped to furnish both his own and the Kathmandu government with a set of alternative proposals. In turn he hoped that these might initiate changes in the structure of Nepalese society and in the nature of Anglo-Nepalese intercourse that would draw off some of the built-up tension. He readily identified that the feudal hierarchy within Himalayan society was organized solely to provide an aristocracy with no other vocation than military command. Meanwhile, the poor subsistence farmers were constantly tempted to abandon the plough in favour of the sword because of the opportunities of booty offered by warfare. In a paper on the Himalayan aboriginals published in 1832 in the *Journal of the Bengal Asiatic Society* Hodgson argued for large and regular inductions of Nepalese warriors to supplement the British ranks and in order to drain off Nepal's surplus troops. Their preoccupation with caste and its attendant taboos about pollution rendered the Hindu sepoys of India of dubious value to the British government. These hill tribes, on the other hand, more or less indifferent to ceremonial law, and more hardy and courageous—facts well attested by the fierce fighting of the Anglo-Nepalese war—would make first-rate troops. For the next 25 years, unfortunately, the British governors, with a suspicion born of ignorance, chose to disregard Hodgson's arguments, and it was only with the Indian Mutiny of 1857 that circumstances forced a reversal of that decision.[120]

In addition to this measure, Hodgson recognized that the martial spirit of the Nepalese had to be redirected into more constructive and profitable channels. He felt that trade, which did not particularly flourish in Nepal, might be made to compete for those energies currently expended on warfare, and by a series of careful enquiries into commercial matters hoped to ensure that it did. His analysis was painstaking: not only did he ascertain which commodities would be suitable for the Nepalese market, but also the merchant's expected return on his goods, details of the levies that would be imposed, the cost of porters, even the price of bullocks for transportation. The purchase price of goods in Calcutta was estimated on items as disparate as carpets, coral, betel nuts, and gun flints. Such thoroughness was indicative of Hodgson's depth of vision, and he was not unrewarded. Within sixty years he lived to see a tenfold growth in Anglo-Nepalese commerce amounting to an annual figure of 33 million rupees.[120]

However, it was not simply a case of pointing the direction to the Nepalese and watching his proposals flourish. He had also to win the confidence of the Indian traders and encourage them to come to Nepal to do business. These merchants were wary of an unfavourable bias in Nepalese legal matters towards the country's native businessmen. Hodgson therefore set out to investigate the Nepalese judiciary, and to arm Indian merchants with a written code to which they could refer in any legal dispute. With the help of Brahmin pandits whom Hodgson invited to answer a series of 93 questions drawn up by himself, he prepared a document in 1832 which eventually persuaded the Nepalese durbar to take into account the status and rights of foreign merchants in their country.

One final way in which Hodgson hoped to alleviate the conflict between the two sides was the establishment of clearly defined national boundaries. Hitherto both sides had been able to exploit any area of ambiguity for their territorial gain and Hodgson hoped that mutually agreed borders could obviate further dispute. In 1833, under a British chairman, a commission was set up and an acceptable demarcation worked out. However, Bhim Sen Thapa was suspicious that this, together with the legal and commercial agreements, were part of the British efforts to infiltrate Nepalese society. He saw them as a threat to national integrity and hoped to put off indefinitely their implementation. Fortunately for the British, the death of the Queen Regent in 1832 and the destabilizing effect that this had on the durbar, forced Bhim Sen to come out of his corner and secure good relations with the Company to bolster his position at court.

The repercussions of the royal death were far greater than a few commercial agreements with the British resident, however. It brought to a close the long and successful coalition between the Queen Regent and Bhim Sen, and ushered in a period of civil unrest that frequently threatened to spill over into armed conflict with the Company. Throughout his years in power, the chief minister had used his influence to fill vacant political posts with his own kinsmen. This policy, together with his efforts virtually to imprison the young Raja, Rajendra Bikram Shah, in his palace had eventually alienated many different elements in Kathmandu. Chief among these was the senior wife of the young Raja, and her champion, Ranjang, the head of the Pandi family. The Pandis, like Bhim Sen's own family, the Thapas, had once held the leading ministerial positions in the Government until their virtual extermination at the end of the eighteenth century. Bhim Sen had been responsible for this massacre and had authorized the appropriation of their land and property. Now with the old Queen dead, and the present Raja dominated by a wife sympathetic to their claims, the surviving Pandis were set on revenge.

In 1833 at the Pajani, or annual vacating of office, a ceremony which Bhim Sen had previously used to secure his own pre-eminence, the Raja, now in his majority, asked Bhim Sen to lay down the seals of his office. No alternative figure was found to replace him and Bhim Sen was eventually reappointed, but this important symbolic event signalled a new era in Nepalese politics. In 1837 the Pandis, under the leadership of Ranjang, and with the direct connivance of the Raja's first wife, brought charges against Bhim Sen that he had poisoned the previous Raja and his Queen in 1816. The court physicians were cruelly tortured but the inquisitors failed to obtain the evidence to convict Bhim Sen. Not content with his simple incarceration and dispossession, the Pandis continued to persecute and torture the ageing chief minister until, utterly distraught, he took his own life to vindicate his innocence in July 1839.

The wheel of fortune had turned full circle: now it was the turn of Ranjang Pandi to fill the offices of state with his own relatives. The pacific policies of the old ministry were overturned and replaced with a plan to expel the British resident and unite with a number of Indian states and foreign powers inimical to Company design. For three years until the fall of the Pandi faction, the durbar in Kathmandu bustled with envoys either setting out for or returning from the courts of their allies. Afghanistan, Persia, Russia, China, Punjab, Gwalior, Satora, Baroda, Jodhpur, and Jaipur were all drawn into the Nepalese strategy to defeat the East India Company. These preparations for war coincided with the period of Britain's increasing involvement in Afghanistan which severely stretched Company resources and weakened its defences along the northern border with Nepal. When surgeon William Brydon staggered into the British fort at Jallalabad, sole survivor from a force of over 16 000 massacred outside Kabul, it was the first time in Company history that its troops had not returned with the spoils of victory. British military prestige reeled, and this sent a flood of rumours throughout India's bazaars that Company power was on the wane.

During this period of crisis when the East India Company's fortunes seemed to hang in the balance, Brian Hodgson's position as British resident in

Kathmandu was one of utmost importance, and one which demanded great delicacy. In an attempt to steer the situation away from the possibility of war and towards relations more favourable to the Company, Hodgson was slowly entangled in the web of Nepalese court intrigues from which he never really escaped. In order to assess his role in the events between 1839 and 1843, events crucial to the development of Nepalese history, it is necessary to give some details of the political situation in Nepal.

With the death of Queen Lalit Tripuri and the subsequent degradation and suicide of Bhim Sen Thapa, a gap was created in the durbar which no single individual was strong enough to fill. Mathabar Singh Thapa, the charismatic nephew of Bhim Sen and commander of Nepal's eastern army, had been forced into exile at Simla on the death of his famous relative. The Maharaja Rajendra Bikram Shah, having eventually freed himself from the oppressive tutelage of Bhim Sen, now found himself harangued and intimidated by his senior wife, Samrajya Laxmi Devi. This fiery Amazon, wilful and violently independent, was deeply resentful of Company intrusion into Nepalese affairs, and until her death from the dreaded 'awal' in 1841 she remained the bitter opponent of Brian Hodgson. She hoped to bully her vacillating husband into abdication in order that she might rule in his stead through her son, the Crown Prince, Surendra Bikram Shah. To attain her objective she chose as her agent, Ranjang Pandi. This man, incompetent, cruel, and completely out of touch with the developments of the last 30 years, hoped to revert to the 'good old Gurkha policy' of war and expansion that had existed before the Anglo-Nepalese war of 1814. However, he was no Bhim Sen Thapa, and found himself unable, even with the senior Queen's backing, to bend the Maharaja to his designs. Rajendra Bikram Shah, although not the most effective of rulers, was no fool, and endeavoured to keep in his hands control of Nepal's foreign and domestic policy. He was unwilling to cede more than nominal powers to his chief minister, but at the same time unwilling to take the lead himself—thus, perhaps inadvertently, helping to perpetuate a political vortex into which the Company's representative was inelluctably drawn.[144]

Through his long years of residence in Kathmandu and also as a result of his extensive researches into Himalayan affairs, Hodgson had developed a deep understanding and sympathy for the Nepalese people. Meanwhile, his en-quiries into the law and religion of Nepal had won him the respect and friendship of many of the countries leading pandits. His own growing fear of the possibility of war between his adopted and his native countries, together with this position of prestige which he enjoyed among members of the Nepalese durbar, prompted Hodgson to depart from the resident's traditional position of strict neutrality. The occasion of his intervention followed upon two acts of aggression against the East India Company by the Pandi administration in the spring and summer of 1840.

The first, in March 1840, involved a small principality on Nepal's southern border called Ramnagar, which was unjustly claimed by the Nepalese on the death of its ruler. In defiance of existing treaties and before the British could respond, the Pandi administration seized the land. Soon after Company demands for its return, the Nepalese troops in Kathmandu mutinied and threatened to attack the British residency. Hodgson was shaken by this incident and was alarmed to discover that the Pandis had been at the back of it. He roused the Governor-General on the basis of this and the Ramnagar affair to threaten the durbar with military sanctions. Troops were summarily brought up to the border with Nepal, and strong demands made, both for the return of territory and compensation. In both cases the Maharaja acquiesced, punishing the guilty, and handing over 5000 rupees as payment for damage. Hodgson then capped these successes by demanding on behalf of the Governor-General, the dismissal of the Pandi ministry and its replacement with one more acceptable to Company interests. In the winter of 1840 this measure also met with success, and Chautaria Fatteh Jung Shah, a collateral of the royal family, became the new chief minister. To demonstrate the extent of the durbar's change of heart and the sincerity of its good intentions, a document was drawn up and signed by 94 members promising to uphold good relations with the Company. In the spring of 1841 all seemed well.[144]

Having started out from a position of censure of Hodgson's involvement in durbar affairs, Lord Auckland, the Governor-General, by the end of his term of office in February 1842, had come to approve wholeheartedly of his resident's actions. Indeed, for a while it was Hodgson himself, with Auckland's approval, who had determined Company policy towards Nepal. Shortly before his departure for England, the Governor-General wrote to his resident saying that he had 'acted throughout these transactions with a thorough knowledge of the native character and with a degree of skill, prudence and forebearance that is highly creditable to you'. Unfortunately for Hodgson, Auckland's successor did not hold the same views.[127]

Lord Ellenborough, shortly after his assumption of office, had decided to pursue an energetic policy on his own behalf from Calcutta. Residents at native courts were required to act as neutral intermediaries simply administering the decisions taken by central government. As Hodgson became increasingly embroiled in Nepalese affairs in an attempt to prop up the administration of Fatteh Jung Shah, it was inevitable that he would clash with the new self-determined Governor-General. When Ellenborough ordered his resident to return to a position of strict neutrality, Hodgson pleaded to be

allowed to continue his existing policy and support for the new chief minister. However, as Ellenborough rightly pointed out, to become the ally of one
party in the durbar was automatically to become the opponent of all the
others. That Hodgson's active intervention had aroused the suspicion of the
Raja and alienated various sectors of the public there can be no doubt.
Placards had been hoisted above the Tundikhel parade ground denouncing the
Shah administration as traitors for its involvement with the resident. Nor can
it be denied that Hodgson's support for the new ministry had encouraged the
belief in the durbar that no government could survive without British
backing, a misconception that tended to destabilize Nepalese politics.[144]

Ellenborough had decided that, along with his predecessor's policies, his
men would have to go, too, and since Hodgson was excluded from further
service on the plains, he would, in effect, be forced to retire. In fact, he
remained just long enough to see a successful transfer of British policy in
Kathmandu and then departed in December 1843. It is interesting to note that
for all Hodgson's opposition to many of the affairs in the Nepalese court, its
members, when they discovered that he was to be relieved of his post, did
their utmost to change Ellenborough's mind. The Maharaja's own personal
communication with the Governor-General read:

I have been perpetually reflecting upon Mr Hodgson's perfect knowledge of the
customs and institutions of my kingdom and likewise upon his long and zealous, kind
and patient labours in the late troubled times, whereby the designs of evil persons
inimical to both governments, were foiled and peace and true friendship with your
state preserved.
   The more I think upon these invaluable qualifications and exertions the more I am
pained at the idea of his departure.[144]

Such protests on Hodgson's behalf testify to the fund of goodwill and even
friendship which had managed to develop beneath the most strained of international relations. In the final analysis, Hodgson had truly cared about the
well-being of his adopted country, and, when confronted with the possibility
of his departure, the Nepalese people were ready to acknowledge that fact.
   Yet Ellenborough was adamant. On 5 December 1843, in an almost regal
procession, the old 'Hermit of the Himalayas', after 23 years of services in
Nepal, left for the border never to return.

LIFE IN RETIREMENT

Hodgson's departure from Nepal and resignation from the East India Company was by no means his cue to a leisured retirement in England. Less than a
year after his return to the family home in Canterbury found him restless and

depressed. Not a man to enjoy wasting time, he was bored by the round of society meetings that his honorary status obliged him to attend, and by October 1844 he had decided to return to India. The British authorities turned down his request to settle back in Kathmandu but allowed him to live in Darjeeling, a hill town in the eastern Himalayas. Geographically and climatically it was very similar to Nepal and well suited the continuance of his researches.

From an old associate, Herbert Maddock, he bought a bungalow, which he renamed Bryanstone and which overlooked the eastern Himalayas from the slope of a 2600 m ridge. Living here Hodgson saw virtually nobody apart from another old colleague from his Kathmandu days, Dr Campbell, who was superintendent of the town. One notable exception, however, was the famous botanist Joseph Hooker, who stayed with the ex-resident for many months during his expeditions to Sikkim, India, and eastern Nepal. The young botanist's letters testify to Hodgson's unrivalled knowledge of Himalayan subjects, his deep pride, and his characteristic generosity. For Hooker, he was the perfect host and tutor, freely divulging the fruits of his own research and then assisting Hooker's own expeditions with moral and practical support.[128]

It seems that Hodgson found great satisfaction in his own company, yet when the opportunity presented itself he delighted in attending to the well-being of his guests. His friend and later biographer, William Wilson Hunter, recorded this period as one of the happiest in Hodgson's Indian life, particularly after 1853 when he returned from a short trip to Europe with a bride. He and his wife, Ann Scott had met during Hodgson's visit to his sister Fanny in Holland, and they were married soon after. They returned to India and lived in Darjeeling for the next six years.

It was during this later period that Hodgson was entrusted with the education of the heir apparent to the Nepalese throne. Through his contact with the prince, and the boy's father-in-law, Jung Bahadur, the all-powerful prime minister in Nepal, Hodgson was able to play a significant part in pressing the British authorities to accept Nepalese military aid during the Indian Mutiny of 1857. After a full enquiry the Governor-General agreed to the proposals and thus opened the door to the permanent Gurkha recruitment that Hodgson had advocated over 25 years earlier.

Soon after the Company's greatest upheaval in its Indian history, Brian Hodgson faced his own personal crisis. This time his wife's health had succumbed to the Indian climate, while his father, then a widower, pleaded with his only surviving son to return to England. A dutiful son and husband, Hodgson abandoned his old intention to prepare a detailed work on Nepalese history and sailed for England in 1858. The materials which he acquired, though insufficient to complete the work, filled several large trunks, and were donated to the Indian Office Library in 1864.

It is remarkable to think that Hodgson, having suffered so terribly with ill health in India and Nepal, should have enjoyed so long and healthy and active an old age when he returned to England. Yet his family had been noted for its longevity, and Hodgson, whose own lifespan was equal almost to that of the nineteenth century, was the longest-lived of them all. He spent the next 36 years as a retired country squire, first in Dursley, Gloucestershire, and then later at Alderley in the Cotswolds. Although he had withdrawn from public scholarship he retained an active interest in his former intellectual pursuits, publishing three volumes of his collected essays, one in 1874, then two more in 1880. A staunch Liberal amongst a host of largely Tory squires, Hodgson frequently received eminent scholars such as Max Müller and Joseph Hooker. Until his very latest years he and his second wife, Susan Townsend, whom he married in 1870, would spend the summer months in London and then the coldest winter months at their villa in Mentone on the French Riviera.

Brian Hodgson died in May, 1894, a remarkable man who shared so many of his century's best qualities, such as its honesty, its restless curiosity, and its energy. Yet he also anticipated the values of a later age: particularly modern in character was his emotional and intellectual sympathy for foreign customs and foreign people; then its political corollary, a belief in the ways of peace. While there is no doubt that he was the product of a previous age, that he shared so many of our own values—ideals believed to be an advance on those of our ancestors—is a measure of his modernity and his vision.

2

A Renaissance man

BRIAN HODGSON was born in an age when, almost as if for the first time, men opened their eyes to the world around them and were fascinated by what they found. He is an excellent example of that type who felt it almost a religious duty to advance the bounds of knowledge, seizing every opportunity to investigate and make sense of the things that he saw. In our own age of ever-increasing specialization where, irrespective of whether one is a scientist or a student of the arts, the tendency is to concentrate on one small portion of a whole discipline, it is remarkable to encounter a person so broadly gifted as he was. Even if one documented his efforts in the fields of Buddhism, anthropology, and the natural sciences, the full range of his talents is not exhausted, since he more than dabbled in a number of other subjects and often made important contributions to them. Equally, any attempt to pigeon-hole Hodgson's studies into a number of different compartments is likely to give a false notion of his approach to study. For, quite simply, everything seems to have interested him—from the cultivation of camellia bushes to the pre-historical migrations of the human race. It is therefore more correct to see his scholarship as his efforts to recreate a whole panorama rather than a series of separate and narrow vistas.

The sheer volume of his output is no less remarkable than the breadth of his focus. Between the years 1828 to 1857 he wrote over 200 papers, an average of one every seven weeks for 30 years. Apart from the large number on natural history dealt with in later chapters, he produced a further 60 on other subjects, principally Buddhism and anthropology. All of his work was done as an amateur, much of it on ground previously untouched by others. In addition to his researches, Hodgson shared to the full the nineteenth-century obsession with collecting, and amassed two large collections of religious manuscripts and natural historical specimens, principally birds and mammals. It is perhaps for these, which he generously donated to museums and institutions throughout Europe, that he is best remembered.

As early as 1821, soon after his arrival in Nepal, Hodgson started to acquire Sanskrit manuscripts, spending large portions of his salary on their purchase. During the course of his inquiries into the subject he met and befriended Amrita Nanda, a distinguished pandit living in Patan a town close to

Kathmandu. As their friendship grew, the old Buddhist began to procure for Hodgson an increasing number of those manuscripts whose whereabouts he could trace. Then in 1824, in order to furnish himself with some understanding of his newly discovered texts, Hodgson drew up a list of questions which he put to his religious associate. The answers which Hodgson wrote down formed part of the first direct and accurate account of Buddhism that the Western world possessed. However, it was not until after another five years of patient sifting and revising that Hodgson was fully satisfied with the results. Limitations of space disallow a full analysis of these discoveries, but they were published in 1828 in the *Transactions of the Royal Asiatic Society* and were received in Europe as revelatory.

Today, Hodgson's sketch of Buddhism is no longer essential to an understanding of the subject; it has been described as 'cumbersome, often misleading and incomplete' yet before Hodgson, as he is often anxious to point out, there had been no thorough description or appraisal of the religion, and his work was the first building block upon which Western understanding could be established.[148] Max Müller, the famous philologist and orientalist of the nineteenth century, said that, prior to Hodgson's work, 'information on Buddhism had been derived at random from China and other countries far from India and no hope was entertained that the originals of the various translations existing in these countries would ever be discovered'.[127]

The romantic Csoma de Körös, who with little money and no more possessions than he could carry, had set out on foot from his home in Hungary in search of his people's true origins, had at this time also been working on the Buddhism of Tibet and Nepal. Although both he and Hodgson had made their researches contemporaneously neither knew of the other's existence, and the latter had the good fortune to publish his findings first. De Körös praised highly his contemporary's work, saying that it formed 'a wonderful combination of knowledge on a new subject with the deepest philosophical speculations', and that it would 'astonish the people of Europe'. Hodgson continued correspondence with de Körös and eventually supplied the manuscripts that formed the basis of some of de Körös' own work on Buddhism.[127]

Hodgson's discovery of the Sanskrit texts also helped to fuel a vigorous and ultimately fruitful debate that continued for many years between those studying the Nepalese works and those Buddhist scholars working in Sri Lanka (or Ceylon, as it was then called). The latter maintained that the first written account of Sakyamuni Buddha's teaching, drawn up nearly 350 years after his death in three huge convocations, was made in the Pali or Prakrit language. They claimed that the canon of works which they had found in Sri Lanka, and which dated from between 102–75 BC in Pitakataya, were transcriptions of those earliest works. Hodgson disagreed. Though he conceded that the language used to preach Buddhism to the common people would most likely have been the vernacular Pali, it did not follow that the same tongue was used to develop abstruse theological arguments. On the contrary, he believed that the early Buddhists would have used the language best suited to such a task, one that had been the medium of philosophical writings for hundreds of years, which was Sanskrit. Comparisons of Chinese, Burmese, and Siamese translations with the earlier Sanskrit and Pali versions, tended to reveal that the later manuscripts bore greater similarity to the Sanskrit, suggesting that this had been the original source of the translations. The discovery of some Chinese works prepared in the first century AD from earlier Sanskrit texts put the issue beyond question. For those originals to have acquired the antiquity and religious authority necessary to be accepted as the scriptures of the Buddhists and therefore worthy of the translation by the people of China, they must have existed two or three centuries before that time—a date long anterior to the Pali manuscripts found in Sri Lanka. Hodgson's arguments had carried the day.[134]

As a result of Hodgson's ascetic life-style, his high reputation with the Nepalese pandits, and his knowledge and interest in the literature and faith of the Himalayan people, he was presented with several important gifts. One of these, from Tibet's own Dalai Lama, was a set of two great encyclopaedias known as the Kahgyur and Stangyur. These were made up of 334 bulky volumes printed with wooden blocks on Tibetan paper and comprised a whole cycle of Tibetan sacred literature. The Kahgyur, whose name means 'translation of commandments', consisted of a hundred volumes. These were supposedly translations of texts taken from between the seventh and thirteenth centuries AD, largely written in the Magadha language. The Stangyur was a miscellaneous collection of over 225 volumes on various subjects, largely written by Indian pandits and some learned Tibetans from the first century of Buddhism's introduction into Tibet, in the seventh century AD. Hodgson eventually received two sets, one of which he passed to the Bengal Asiatic Society in 1829; the other given to him by the Dalai Lama in 1835 went to the Company directors in 1838. That Russia's Tsar had to pay £2000 for half the series is indicative of Hodgson's munificence in these donations. Other presentations he made were a number of Buddhist drawings to the Institute of Paris and a large collection of Sanskrit, Persian, and Newari manuscripts to the Secretary of State for India in 1864. In the same year he generously endowed the British Museum with a Tibetan translation of the *Prajna paramita*, an enormous religious work consisting of five volumes and 100 000 verses.

As repositories for his collection of more than 400 Buddhist texts, some of which dated from the eleventh and twelfth century AD, Hodgson selected six libraries in England, France, and India. Those 160 manuscripts which he deposited in institutions in Calcutta, India, formed the basis of Csoma de Körös's subsequent work on Buddhism, and were later catalogued by Dr Rajendra Lal Mitra in his *Sanskrit Buddhist Literature of Nepal*. Those he delivered to France, almost 150 texts, helped to establish the first professorial chair for Sanskrit in Europe, and were used by its first incumbent, Eugene Burnouf, to write his monumental *Introduction a l'Histoire du Buddhisme India*. Burnouf acknowledged Hodgson as the 'founder of the true study of Buddhism', and dedicated to him his final and posthumously published work, *Le Lotus de la Bonne Loi.*

Over a period of ten years, Hodgson continued to contribute his own essays on Buddhism to the Bengal Asiatic Society, and supplied other shorter pieces on Buddhist inscriptions and monuments. In 1841 these were published collectively in a work entitled *Illustrations of the Literature and Religion of the Buddhists*, and were republished by Trübner in 1874 as *Essays on the Language, Literature and Religion of Nepal and Tibet.*

Albrecht Weber, an eminent Buddhist scholar of the nineteenth century and correspondent of Hodgson's, wrote in one of his letters, 'you may be sure the path that you have opened that its importance will be more and more acknowledged each year'. On the continent, this was certainly the case. In 1838 French academics made him a Chevalier of the Legion of Honour and the Société Asiatique awarded him a gold medal for his services. In 1844 he was also made a corresponding member of the French Institute and an honorary member of the German Oriental Society. While other societies throughout Europe elected him to honorary status, his own countrymen were far more chary in recognizing Hodgson's achievements, however—although, as his biographer William Hunter pointed out, with only a handful of exceptions, almost all original work on northern Buddhist manuscripts in France, Great Britain, and India during the half-century after Hodgson's discoveries had been based on his collections. One of those exceptions, Professor Bendall, who had researched Sanskrit manuscripts at Cambridge, acknowledged Hodgson as the 'greatest and least thanked of our Indian residents'. Yet it was not until 1889, when Hodgson was approaching his ninetieth year, that the University of Oxford conferred upon the 'father of modern critical study of Buddhist doctrine' a Doctorate of Canon Law.[127]

It is not surprising, given the broad sweep of Hodgson's studies, that his contemporaries had some difficulty in isolating the man's principal contribution. Entries or obituaries in encyclopaedias, journals, and dictionaries giving a brief summary of his life, frequently focused on one or another of his interests, probably reflecting those of the author more than the subject himself. Hodgson's own longevity might have played a part in obscuring his exact contribution, since the 36 years that had elapsed between the period of his scholarship and the date of his death tended to leave few people who could recall what he had actually done.

Yet all his immediate colleagues in each respective discipline joined in a chorus of approval of his achievements, choosing to overlook the minor faults of this self-taught scholar in favour of the mass of new information he provided. Hodgson had neither library nor fellow enthusiast to encourage him or point a direction, yet he seems to have made an asset of his disadvantages. Isolated from European society and often confined by ill-health, he took advantage of his solitude in Nepal to devote himself uninterruptedly to study. His arguments and speculations were always highly original and accurate, drawn from his personal observations of the subject in question.

When Hodgson started his ethnological studies he was all the more a pioneer for embarking upon a discipline still very much in its infancy. His work should therefore be seen within the context of his early Victorian approach, which was largely aimed at tracing the historical developments of the human race. Anthropologists, or ethnographers, as they were called, applied to the study of mankind the model of investigation that naturalists such as Darwin and his contemporaries had used in studying the animal world.

They believed that man had passed through several different stages during the course of his evolution and that these stages were historically pre-ordained. Not only could they predict the path that less-developed races must take towards progress, but, in studying these people, modern man could look through a window into his own primitive past; and the more undeveloped the tribe or ethnic group the further nineteenth-century man could delve into his own beginnings. The existence of small isolated tribes was particularly fascinating to them, since they thought that these were the survivors, the living relics, of man's earliest ancestors. When pieced together in a chain, these different groups formed a complete picture of man's successful advance from cave dweller to nineteenth-century industrialist.

Hodgson's own enquiries fell largely within this pattern, two basic pre-occupations emerging from his work. The first was the geographical and historical origins of India's indigenous population. The second was the interaction between those aboriginal people and the later more successful Aryan Hindu invaders.

His early studies of the Himalayan and sub-Himalayan tribes led him to conclude that 'the mountain people were of one common stock'. That stock,

or family, he called Turanian, to designate their approximate land or origin
—Tur, or modern-day Turkestan in Soviet Central Asia. During the course of
his 25 years of research the substance of his views changed very little, he
merely widened the scope of his research to include greater and greater
numbers of the non-Aryan people. In 1833 while he was still in Nepal and
studying that country's military tribes he limited the descendants of the
Turanian family to those people occupying the hills and mountains of the
Himalayas. By 1848, however, when he was still writing up his work on the
tribes of northern Tibet, he had developed a 'conviction that the Indo-
Chinese, the Chinese Tibetan, and Altaians had been too broadly contra-
distinguished, and that they form in fact but one great ethnic family, which
moreover includes what are usually called the Tamulian or Dravidian, and
the Kol and Munda elements of the Indian Population'.[120]

Since Himalayan tribes had remained in areas largely undisturbed by the
Hindu Aryans, their language and customs has remained more or less intact.
It was, therefore, possible to trace any racial affinities these people might have
with those further north. This was not the case in the Indo-Gangetic plains
further south, however, for in the second millenium BC the original Dravidian
inhabitants, themselves of Turanian stock, had been affected by the succes-
sive invasions of Aryan Hindus from the north-west. Where they had not been
displaced by the stronger and more successful Hindus, their language and
culture had been absorbed into the latter's social structure. Those who had
withdrawn to the Deccan Plateau further south and had successfully resisted
further Aryan advances, had retained their racial and cultural identity. Those
tribes which had been fragmented during the course of their retreat into
jungle and other unhealthy environments had become extinct.

The central Himalayas had not entirely escaped the effects of such a process
whereby a single dominant group had affected the surrounding tribes.
Hodgson devoted several papers to the Khas military tribe, who themselves
had been 'Indianized' by the influence of the Hindus, and had then, in turn,
acted upon their neighbours. Tribes like the Kusunda and Chepang, in the
struggle for land, had been broken up by their more successful competitors.
They were exactly the sort of relic group in which early ethnographers took
delight, and Hodgson recorded their existence in a paper published in 1857,
only a year before his final departure from India. He concluded it by saying,
'the lapse of a few generations will probably see the total extinction of the
Chepang and Kusunda, and therefore I apprehend that the traces now saved
. . . will be deemed very precious by all real students of ethnology . . . as
proving how a tribe may be dislocated during the great transitional eras of
society'.[121]

Most of Hodgson's ethnological theories were hammered out on an anvil of linguistic analysis, his work being as valuable to the philologist as to the anthropologist. He was himself a gifted linguist and had a thorough grasp not only of widely-spoken native languages such as Urdu and Hindu, but also the more restricted hill tongues and dialects. Hodgson's papers on the various tribes of the Himalayas and other hill ranges often consisted of long and exhaustive comparative vocabularies. His claims for a common family for almost the entire non-Aryan races of India were based upon a painstaking examination of lingual affinities. He thought that where an indigenous language had escaped the dominant influence of the Aryan Hindu, its true racial and linguistic origins—that is, Turanian—would be apparent.

To read Hodgson's various essays is to be left with the impression of an intellect struggling to grasp the whole of an issue. He was never content to work blindly in some narrow corner, but wished to survey and speculate on the entire picture, even if some of the details were missing. In 1856 in a paper on the Aborigines of the Nilgiris, he wrote: 'I am decidedly of the opinion that the true relations of the most shifting and erratic, the most anciently and widely dispersed, branch of the human family cannot reasonably be investigated upon a contracted scale, while the subject is so vast that one must need search for a feasible means of grasping it in significant amplitude.'[121]

Later shifts in the mainstream of anthropological thought have largely rendered Hodgson's theories redundant or inaccurate. They were no longer concerned with fitting a social group such as the Chepang and Kusunda tribes into some hypothetical time-scale and building up a chronological picture of man's development. They felt it more important to know, not that a particular ritual or custom survived, but why it existed, and what function it served in the social structure of a particular ethnic group. There were also changes in the methods of recording anthropological data: field researchers now lived closely with their study group, or on an equal footing, participating as much as possible in its various activities. Political officers such as Hodgson, who indulged in 'spare time' anthropology, often saw their subjects as inferior beings, biological specimens to be treated as one treats a rare butterfly or beetle. Many of Hodgson's ethnographic accounts include lengthy and detailed anatomical studies, demonstrating an assumption that rare tribes were living specimens, waiting to be catalogued. They believed that human skull size and shape was a means of classification, and Hodgson's essays are peppered with such biometric data. Indeed, one of his natural historical donations to the British Museum in London was a collection of human skulls.

However, this is not to undermine Hodgson's achievements. Although his theory about the common origins of the Mongol, Turkic, and Dravidian people of Asia and the Indian subcontinent is now largely seen as over-generalized and inaccurate, his extensive collection of vocabularies, and his often sympathetic records of the social customs of tribes such as the Koch, Bodo, and Dhimal of north-east India are valuable sources of general information. Certainly in his day Hodgson's work on ethnology was held in very high esteem. A contemporary thought him 'the highest living authority and best informant on ethnology of native races of India'. He was a regular correspondent with the leading philologists and ethnologists of his day such as Latham, Schleicher, Lassen, Max Müller, and Dr Barnard Davis. The first of these said of Hodgson that 'he will do more to clear away the rubbish and restore to the lost annals of the Gangetic Valley than Lassen, with all his erudition and talent for historic research'; even Lassen himself had 'a high sense of admiration' for Hodgson's services.[127]

In 1868, Hodgson's biographer, William Wilson Hunter, published *A Comparative Dictionary of the non-Aryan languages of India and High Asia*. In the introduction he acknowledged his debt to the old resident, who had supplied almost all the basic data and had put at the author's disposal two large trunks of his own researches and manuscripts. Hunter said, 'in some respects I look upon myself as the editor of materials collected by him (Hodgson) rather than as the author of an original work'. A hundred years after Hodgson had written his first anthropological paper, Sir George Grierson published the enormous *Linguistic Survey of India* in 1927. In the introduction he wrote, 'in 1828 we first meet one name that overshadows all the rest—that of Brian Houghton Hodgson . . . his writings contain a mass of evidence on the aboriginal languages of India that has never been superseded'. Closer to our own time, recent works by another expert on Himalayan anthropology, Christopher von Fürer-Haimendorf, frequently cite Hodgson's various papers. While it may not be said that Hodgson's work on Buddhism enjoys a living presence in the references, bibliographies, and footnotes of contemporary authors, this is not true of his ethnological studies. That some of them were written over 150 years ago is testimony to how well the self-taught scholar had completed his task.[14]

It is not surprising, given Hodgson's interest in Himalayan fauna and the various hill tribes of the region, that he also gave thought to the structure of the Himalayan mountains themselves. Until his time, it had been suggested that the great peaks of the Himalayas intersected the river basins of the area's main drainage systems. Hodgson argued otherwise. The fact that the multitude of small feeder streams, coming from a maze of ridges and hills, converged to form three great rivers—the Karnali, the Gandak, and the Kosi—suggested to him that the great peaks formed the boundary of these drainage systems. It was the immense height of these mountains and the ridges that ran from them in a southerly direction that determined the direction of their flow, and overruled any other effect the intervening hills might have upon them. Confirmation of his arguments was provided by his discovery that at the very highest peaks, with their correspondingly high longitudinal spurs, the major river systems reached a point of intersection. For example, at Dhaulagiri, a Nepalese mountain of over 8000 m, the feeders of the Gandak flowed off its slopes to the east, and those of the Karnali to the west.

While his geographical studies shared a close relationship to his other interests, he also wrote a completely different type of paper of a more practical nature. Like so much of his work, these papers were often highly original and, as with his advocacy of Gurkha recruitment into the British army, showed tremendous foresight. Unfortunately, they were similarly ignored by his peers, often without justification. Most valuable of all were his works on the preparation or cultivation of Nepalese paper, silk, Tibetan wool, hemp, and tea. In 1857, just before his return to England, he again announced the suitability of the Himalayas for the production of tea—a truth now widely adopted and one playing a significant part in the region's economy. Yet Hodgson had first known and stated it over a quarter of a century before.

Admittedly some of his more adventurous suggestions still sound fairly radical today, such as his promotion of European colonization of the Himalayas. It is perhaps better not to contemplate the consequences of such a project had it ever been mounted, yet the logic of his argument, political considerations notwithstanding, is obvious. There was a large surplus and poor agrarian population in the remote districts of Scotland and Ireland, and there was a large 'surplus' area of fertile land in the Himalayas. It was a simple case of transferring them one to another, and within a number of years the British in India would have a loyal and locally-based militia to defend its northern border. The proposal might seem preposterous today, yet within the intellectual and political context of Hodgson's time he could probably see little to stand in its way.

Perhaps of a more acceptable nature were his papers on Nepalese and trans-Himalayan trade, whereby Hodgson hoped to encourage the traffic between the plains of India and Tibet. Again, one can see the naïve logic of his argument. The Russians, Britain's military and commercial opponents, enjoyed a considerable trade with Tibet and western China in spite of large geographical obstacles. The Company, with the variety and high quality of its produce and its easier access to the region, might easily win for itself the central Asian market. At the same time, Hodgson saw this commercial exchange as helping to draw off any military tension that might exist on India's northern border. Other papers of a similar nature were his topographical descriptions, which informed those with a vested interest the distance and nature of routes between Kathmandu and Darjeeling, Kathmandu and Peking, or Sikkim and Kumaon.

However, Hodgson's most important political contribution, and one which clearly demonstrates the absence of nineteenth-century prejudice in his thinking as well as his philanthropic attitude towards his fellow man, was his devotion to the cause of vernacular education. Consonant with the colonial ethic, the British in India felt it their responsibility to activate the spiritual and intellectual regeneration of Indians by means of a national educational system. Equally consistent with the chauvinism of such a political system they chose as the vehicle for that education the English language. This had been decided in 1835 and had been championed by the powerful advocacy of Lord Macaulay, the Chairman of the education committee for the East India Company. Strangely, at the time, the major opponent of the Anglicists was not Hodgson but those who favoured the classical languages of the Orient —Sanskrit, Persian, and Arabic. Hodgson's was a third and unheeded via media.

He was not perturbed. A series of letters written between 1835 and 1837 and published in book form in the later year, put forward very powerfully and clearly the case of the vernacular languages. He first attacked the fallacy of the superiority of English. He conceded its importance as the technical language of the new sciences, but in every other respect the vernaculars with which Hodgson was fully conversant were sufficiently endowed with dictionaries and grammars to become the medium of national education. As English drew upon Latin and Greek, so could the vernaculars draw on Sanskrit and Arabic for the formation of new words. It was not true, as the Anglicists contended, that the vernaculars were a chaos of dialects each the competitor of the other: the vast majority of Indians spoke one of three languages—Hindu, Urdu, and Bengali. Although Hodgson gave careful consideration to the charges and claims of the Anglicists regarding the supremacy of English, the whole question of which language would be most effective for national regeneration was for him a red herring. In people so habit-bound as the Indians, how could one possibly uproot the deepest of national habits—the language itself—and replace it with one completely alien? The only possible vehicle for national education had to be the vernaculars.

On this argument he might have rested his case, but he went on. The Anglicists believed in a trickle-down theory of national regeneration: once the Indians privileged enough to receive an education were set free into the world they would act as well-springs of English thought and philosophy, irrigating the vernacular language of others with fresh knowledge. Hodgson believed otherwise: 'Anglicisation will help to widen the existing lamentable gulf that divides us from the mass of the people and put into the hands of the few among themselves an exclusive and dangerous knowledge.' Rather than use their education for the good of all, the new elite would exploit it for their own selfish ends. In his arguments Hodgson displayed a strong anti-Brahmanical, anti-caste egalitarianism, referring to the monopoly of knowledge enjoyed by India's priestly classes as the 'special curse which hath blighted the fairest portion of Asia from time immemorial'. Use of the vernacular would be an insurance against the 'ancient curse' of exclusive knowledge and would restore self-confidence to Indians by forcing the English to learn and propagate Hindu and Urdu. Rather than dividing society the vernaculars would become a bond between ruler and ruled.[121]

Hodgson did not, as so many parties do, simply erect the case for the vernaculars by dismantling the claims of Anglicists. English should be taught, both to those who could afford it and those educated at public expense, where it, along with the modern sciences, would act as an effective foil to the abuses of knowledge practised by the Brahmins. Indoctrinated with the modern scientific spirit, Hodgson believed that the students would not lapse back upon old superstitious ways. His programme for vernacular education consisted of a three-point charter: first, to win government support for the native languages; secondly, to establish a normal college which would train the vernacular schoolmaster; and they in turn would effect the third part of his plan—the development and refinement of vernacular textbooks.

In the early part of the nineteenth century there had been some support for the vernacular cause, largely in schools taught by missionaries. Yet the education committee's decision of 1835 ended all aid to such establishments. Hodgson's entry into the debate soon after, and the publication of his letters in book form in 1837 ensured that the vernacular voice was heard once again. His arguments, together with an offer of 5000 rupees for the establishment of his normal college, began to acquire support. In 1854 under Lord Dalhousie, the education committee rescinded its original decision and adopted 'a scheme for education for all India' through the medium of the vernacular languages. However, it was not until 1882 that the committee reorganized education along the lines proposed by Hodgson. Yet even then, as his biographer Hunter testifies, the old resident had not lost any of his enthusiasm for the issue. In the final year of his long life, 3 500 000 out of 4 000 000 pupils in Indian schools and colleges recognized by the state were receiving education entirely in the vernaculars.

3

The great discoverer

IN a similar way to Hodgson's anthropological work, his activities as a
naturalist in the Himalayas coincided with the rapid development of this
science in Europe. Initiated as it was in the eighteenth century by early
natural historians such as Linnaeus and Lamarck, Hodgson was part of a
second generation who had begun to discover and systematize the flora and
fauna outside the confines of Europe. Hodgson's contemporaries, familiar
names such as Charles Darwin, Thomas Huxley, Alfred Russel Wallace, and
Joseph Hooker, were beginning to expound their theories of evolution,
zoogeography, and classification; while artists such as John Jacques Audubon
in the USA, and England's own John Gould, had helped popularize an interest
in natural history with their volumes of colour paintings. By 1834 Gould had
published *A Century of Birds from the Himalayan Mountains*,[15] and three
years later brought out his monumental, five-volume *The Birds of Europe*. To
both these works Hodgson had subscribed. The following year, in 1838,
Audubon published his *magnum opus*, *The Birds of North America*.

That Hodgson was an associate—though not always on amicable terms
—with some of these men, is a clear demonstration of his eminence in this
particular field. Hooker had lived with Hodgson for several months at the
latter's home in Darjeeling and thereafter remained a lifelong friend and
correspondent. Hodgson also came into contact with Gould and Darwin,
making available to the latter his notes on the Tibetan mastiff, and Himalayan cattle, sheep, and goats, to assist Darwin in his work on *The Variation
of Animals and Plants under Domestication*. This association with the
leading natural scientists of his day is all the more remarkable when one
considers the difference in the respective positions of these men. Huxley and
Darwin were professional naturalists, while the former also enjoyed the
prestige and privileges that an academic institution could bring to bear on his
behalf. Hodgson, by contrast, was a civil servant, and his researches, as we
have seen, were squeezed in amongst a host of other duties and interests. He
had very little or no education in natural history, while his position in
Kathmandu cut him off from the rest of European society, leaving him
without the stimulating contact of fellow enthusiasts or convenient access to
a well-stocked library.

These difficulties notwithstanding, Hodgson's achievements in the field of natural history are extensive. Although three Europeans who were interested in the subject had visited Nepal before Hodgson—Colonel W. J. Fitzpatrick in 1793, Francis Buchanan-Hamilton from 1802 to 1803, and Nathaniel Wallich in 1817—their contributions are minor compared with his. Hodgson's detailed observations laid the foundations of Himalayan faunal research and provided almost all of the information available on Nepalese wildlife until this century; even today his work on Nepalese mammals, birds, amphibians, and reptiles remains the backbone of our current knowledge. His work on Tibetan fauna was also significant, since he was the first scientific collector of birds and probably also of mammals in the country. His contribution falls into three categories: his extensive collection of vertebrate specimens; his published papers together with a mass of unpublished notes; and finally, a magnificent collection of water colour paintings.

There is no doubt, however, that Hodgson's principal contribution in the field of natural history was his ornithological work. A total bird collection of some 9512 specimens was one of the largest single collections made in Asia and consisted of 672 species of which over 124 were previously unknown to science. Although he is credited with 80 of the latter, the remainder was described by others, mainly Gould, Edward Blyth—the curator for the Asiatic Society of Bengal, and the Gray brothers—curators at the British Museum. Hodgson's work on Nepalese birds meanwhile is unrivalled: only 59 species of the total collection originated from beyond Nepal's borders; the rest represents almost three-quarters of the country's present total.

Testimony to Hodgson's dynamism as a collector is the large number of animal species, mostly mammals and birds, which bears his name. Amongst the English titles of Himalayan birds one finds Hodgson's Frogmouth, Hodgson's Hawk-Cuckoo, Hodgson's Pipit, Hodgson's Redstart, and Hodgson's Bushchat. Equally, in scientific nomenclature one encounters *Columba hodgsonii*, *Batrachostomus hodgsoni* (Plate 1), *Anthus hodgsoni*, *Phoenicurus hodgsoni*, *Prinia hodgsonii*, *Abroscopus hodgsoni*, and *Ficedula hodgsonii*. Even a bird genus—*Hodgsonius* (Plate 2)—is called after him, an honour few naturalists can claim. The names of several mammals also testify to his involvement in their discovery, such as *Hystrix hodgsoni*, *Pantholops hodgsoni*, Hodgson's Flying Squirrel, and Hodgson's Bat. Then, finally, there is a single reptile, a Colubrid snake *Elaphe hodgsoni*, that bears the name of this prolific collector.

Owing to Nepalese restrictions on the movements of British personnel in the residency, most of Hodgson's specimens were collected for him by a team of trappers. Virtually nothing is known about these men except for one,

Chebu Lama, who provided him with some Tibetan mammals from Lhasa.[123] Unfortunately, malpractice on the part of the curators of the British Museum, the Gray brothers, has largely obscured the precise origins of many of the bird skins. Apparently Hodgson had two systems of labelling his specimens: one indicating its name, locality, and date of collection written in Nepali, most of which the Grays deliberately removed and destroyed, the other showing a personal reference number. Each species, subspecies, sex, and often immatures if their physical appearance was sufficiently different, received one of these numbers, and since the Nepali labels were destroyed, they are important as an aid to matching up the individual skins with the Grays' catalogues of the collections[17,18] and Hodgson's own numerical references on the paintings.

The unpublished notes on the back of the paintings are now the major source of information on the origins of the specimens.[122,123] Hodgson frequently refers to those taken in the Kathmandu Valley and the encircling hills in these notes. There were a few skins obtained from Bhimpedi, Mukwanpur, and Hitauda, all villages *en route* from India to the Nepalese capital. However, apart from Gosainkund, a mountainous area to the north of Kathmandu in the present Langtang National Park, Hodgson's trappers seem never to have visited places above 3050 m, nor did they acquire specimens typical of trans-Himalayan districts in Nepal like Dolpo, Mustang, and Thakkhola. Similarly, only three bird species were collected from western Nepal (all pheasants): the Cheer *Catreus wallichii*, the Koklass *Pucrasia macrolopha*, and the Western Tragopan *Tragopan melanocephalus*, the last probably originating from an area west of the border.

It is remarkable to think that in over 20 years of residence in the country, because of the court restrictions on his movements, Hodgson's personal travels were limited almost completely to the 518 km^2 of the Kathmandu Valley. The only exceptions to this were his visits to Koulia, now called Rani Pauwa, the site of the residency bungalow about 22 km from the capital. In his unpublished notes Hodgson often refers to birds and nests that he found at Jahar Powah, a hill on which the bungalow probably stood.[122] There was also Nayakote, a residence of the Nepalese maharaja that he visited when he accompanied the royal family there on the 30-km journey.[120] Then, in 1840, after almost 20 years of confinement, curiosity seems to have got the better of Hodgson and he 'took French leave from the Durbar one day and under pretence of shooting in the Valley, posted off expedite fashion to Doomja', a village in the Sun Kosi valley east of Kathmandu. Unfortunately, this secret expedition was quickly brought to an end by the local people who ordered him back to the city.[78] Owing to these constraints on his movements, the localities of many skins collected outside the Valley were only recorded according to a broad geographical division of the Himalayas, since the exact position of their capture was unknown to him. The great detail concerning those taken in the Valley itself, however, suggests tht Hodgson himself was their collector.

While it is most likely that the species he acquired whilst living in Nepal, and which he claimed came from that country, actually did so, there is a small number of species where this might not be the case. Four birds, the Great Indian Bustard *Ardeotis nigriceps*, Painted Sandgrouse *Pterocles indicus*, Chestnut-bellied Sandgrouse *P. exustus*, and Black-bellied Sandgrouse *P. orientalis*, are all lowland species and may have been collected at a point south of the Nepalese border. Suitable habitat for them does not exist today and it is unlikely that it did then; certainly there have been no subsequent records from Nepal.

If the origins of some of Hodgson's specimens collected while he lived in Nepal seem rather hazy, trying to establish the exact locality of those taken during his later residence in Darjeeling is positively confusing. Darjeeling, a hill town in the eastern Himalayas, lies only a day's walk from the Indian border with both Sikkim and Nepal. Since these borders run through largely undifferentiated hill forest, and since it seems likely that many of Hodgson's specimens were taken by local trappers who could easily have wandered from one country to the next without realizing it, it is difficult to establish from which country they came. Both the catalogues of the British Museum drawn up by the Grays, and those of the East India Company[125] to which he also gave specimens, state that Hodgson's donations originated in Nepal, Sikkim, and Tibet. However, Hodgson himself, in an annotated copy of the second British Museum catalogue, deleted Nepal as a source of these later donations. It therefore seems likely that ten bird species claimed for that country which have not been confirmed by later sightings, such as the Red-faced Liocichla *Liocichla phoenicea* (Plate 3) and the Collared Treepie *Dendrocitta frontalis* (Plate 4), are possibly wrongly listed for Nepal in current literature. All belong to the forests of the eastern Himalayas, and if they ever did occur in Nepal it must have been in the far east.

The later collection included 52 species from beyond Nepal's eastern border together with another seven from Tibet. These latter birds and 33 species of Tibetan mammals all originated from Utsang province which lay in south-east Tibet just north of Sikkim.[111] Some of these additional species were particularly interesting, such as the Orange-rumped Honeyguide *Indicator xanthonotus*. This unobtrusive and drab green bird was known by the Chinese as a spiritual sparrow because of its elusive behaviour. Like the rest of its family it is exceptional, feeding on wax, a substance all other birds find indigestible. Hodgson also collected a subspecies, the Lesser Rufous-headed Parrotbill *Paradoxornis atrosuperciliaris oatesi* which has not been reported this century and may well have disappeared due to clearance of its bamboo and tall-grass habitat in the foothills. Two other enigmatic birds which he found were Hodgson's Frogmouth *Batrachostomus hodgsoni* and the Long-billed Wren-Babbler *Rimator malacoptilus* (Plate 5). The former is a large nocturnal bird, closely related to nightjars but with an owl-like appearance, whose habits even today are little known. The latter, a small ground-dwelling babbler with an abnormally long decurved bill, is renowned as a great skulker and for its ability to avoid detection.

In addition to the separate collections made in Nepal and Darjeeling, Hodgson acquired a small number of bird skins from the Indian state of Bihar. These 301 specimens probably originated from Segouli, a town lying on the old route from India to Kathmandu, and are sometimes mentioned in his unpublished notes. This Bihar collection was prepared differently from his others. The two large collections are unusual in that the birds have their wings outstretched, while the Bihar birds have their wings folded. Hodgson's Himalayan skins were prepared for him by Nepalese whom he must have trained, and are often easily recognizable by their distinctive manner of preparation, as well as by their poor condition.

Hodgson made a gift of all his collections, some 10 500 mammals, birds, reptiles, amphibians, and fish, to the British Museum; the first donation immediately after his departure from Nepal in 1844, and the second when he left the subcontinent for good in 1858. After the museum had selected a series, the duplicate species were distributed to various British and European collections. Meanwhile, the Asiatic Society of Bengal received specimens of 107 bird species and about 12 mammal species.

The first donation to the British Museum, exclusively mammals and birds, was catalogued by the keeper Dr J. E. Gray, and his brother and assistant, G. R. Gray, in 1846.[18] The second, the Darjeeling collection, was completed in 1863, and the second catalogue attempted to produce a combined list of both the first and second Hodgson donations including all his mammals, birds, amphibians, reptiles, and fish.[17] Unfortunately, Hodgson was dissatisfied with this, and a copy which he extensively revised by hand is now held in the library of the Zoological Society of London. He pointed out the omission of 20 mammals and 10 birds, and made numerous changes and additions to localities and references. Gray's alterations of Hodgson's own taxonomy and division of species was also unsatisfactory to him, and in some cases Hodgson's disapproval has since been vindicated.

Gray considered that the two storks, the Greater Adjutant *Leptoptilos dubius* and Lesser Adjutant *L. javanicus* were conspecific, whereas Hodgson had separated them. He also correctly recognized that the Common Hill Partridge *Arborophila torqueola* and Rufous-throated Hill Partridge *A. rufogularis* were distinct species, while the Grays had classed them together. With equal justification Hodgson criticized their putting ten babbler species into the same genus *Leiothrix*: today they are split into the three genera —*Leiothrix*, *Alcippe*, and *Minla*. His severest criticism, however, was of their treatment of his birds of prey. In the amended catalogue he wrote: 'In the raptores, diurnal and nocturnal, two or three of my species are often lumped together and the characters overlooked. Jerdon in his 'Birds of India', with the advantage I enjoy of seeing fresh specimens, has frequently upheld my genera and species disallowed in this and prior catalogues.'

Sadly, Hodgson's specimens also suffered from physical maltreatment at the hands of museum curators, which must at least partly explain their poor condition today. In the introduction to the first catalogue, the Grays stated that 'unfortunately the specimens have been in the country several years and from want of having been opened and examined were not in a very good condition'.[18] Offended by this neglect, Hodgson sent a large number of birds and mammals to the East India Company museum in 1853. Consisting of about 120 bird species and 84 mammals, this collection was similarly abused. In 1874 the Russian traveller Fedchenko was denied access to Hodgson's skins because they were still not unpacked.[11] To cap the irony, the collection was returned to the British Museum (Natural History) in 1881 on the dissolution of the Company museum.

While his work on birds must rank as his main achievement in natural history, Hodgson's collection of 903 specimens of mammals was also important. Of the total 124 species which he found, 87 came from Nepal, constituting about two-thirds of those now recorded in the country; while an extra 37 species came from Tibet and the Indian states east of the Nepalese border. Although 22 species new to science were attributed to Hodgson, there were actually another nine species for which others like Blyth and the Grays took the credit.

Hodgson's 84 specimens of reptiles, amphibians, and fish form the last part of his vast donations. He was the earliest known collector of the first two groups of animals in Nepal and only the second person to gather fish. Unfortunately, the latter, some 20 species, all later catalogued by Albert Günther,[19,20] then curator of herpetology at the British Museum are a complete puzzle. Their place of capture is unknown, while some of the species are inhabitants of salt water.

The reptiles and amphibians are much more straightforward, however. While there is a possibility that some of the specimens originated from further east, it is most likely that the great majority, if not all, came from Nepal. A few of them were described by Cantor in 1839,[9] but it was not until Hodgson finally returned to England in 1858 that he handed over the bulk of the collection to be described and catalogued by Albert Günther.[19,20] Since Hodgson's time, herpetological collecting in Nepal has been sporadic and, indeed, during the first 60 years after he left the country hardly any information was published on these animals.[145] Even today little is known about amphibians (less than any group of vertebrates in the country), and not much more is known about the reptiles.

Hodgson acquired five of the 29 amphibians now recorded in Nepal, including three new species, a toad *Bufo himalayanus*, and two frogs *Rana liebigii* and *Rhacophorus maximus*.[17] He also found 28 species of reptiles, almost half of the country's total.[17] Some, recorded from the foothills and lowlands, are now rare and threatened by habitat destruction, such as the two aquatic turtles *Kachuga dhongoka* and *K. kachuga*, the Claw-tailed Tortoise *Testudo elongata*, the Mud Turtle *Trionyx gangeticus*, and the highly poisonous McClelland's Coral Snake *Calliophis macclellandii*. He also obtained specimens of the Gharial *Gavialis gangeticus*, which once occurred in all major rivers from the Indus in Pakistan to the Irrawaddy in Burma, but now survives only in a few scattered populations in more remote areas such as the Royal Chitwan National Park in Nepal. Four snakes were first discovered by Hodgson and later described by Günther, including his namesake the Hodgson's Racer *Elaphe hodgsoni*, and the most poisonous of Nepalese snakes, the Mountain Pit Viper *Trimeresurus monticola*.

To accompany the various donations of skins, there was also a large osteological collection consisting of birds, mammals, and 95 human skulls which Hodgson presented to the British Museum. In the history of its collections drawn up in 1906, Hodgson's gift was described as the most important donation of its kind from any single individual.[143] This was due not only to its large size, but also to the great number of different types within it and to the fact that it was accompanied by skulls and skeletons. Hodgson himself recognized the importance of his osteological collection and insisted that it be kept with the skins. Up to that date, skulls and skeletons had been thought suitable only for the College of Surgeons, but his donation forced the British Museum (Natural History) to consider making its own osteological collection.

While Hodgson was unrivalled in his time as a collector, he was also a prolific author, with no less than 146 papers on natural history coming from his pen. They were all published in the 32 years between 1826 and 1858, 82 on mammals and 64 on birds. The majority appeared in the *Journal of the Asiatic Society of Bengal* with most of the remainder in the *Proceedings of the Zoological Society of London, Asiatic Researches, Annals and Magazine of Natural History*, and the *Calcutta Journal of Natural History*. They often dealt with the appearance and behaviour of Hodgson's many newly discovered species, but he also wrote on scientific nomenclature and terminology, faunal distribution according to altitude, and was probably the first person to address himself to trans-Himalayan migration. Although some were little more than a few pages, there were some lengthy pieces and many were accompanied by beautiful and accurate illustrations, including a number in colour. His average output of articles for the 32-year period of his studies was a paper every three months. However, his peak was in the three-year period 1835–37 when he produced a little over a paper a month, an outstanding rate for an amateur naturalist.

Hodgson, a corresponding member of the Zoological Society of London since 1832, generously donated to this institution a large collection of original manuscripts relating to his natural history work. This material, some of which has been never published, forms an important addition to our knowledge of Himalayan birds and mammals. The backs of his first collection of paintings were crammed with full descriptions of plumages, furs, colours of bills, legs, feet, and irides, as well as extensive measurements. He also wrote copiously on their breeding behaviour, nests and eggs, habits, altitudes, localities, and dates of collection. Inexplicably, Hodgson usually omitted the year in which his specimens were collected, a fact which compromises the notes' importance. Equally unfortunate is the addition of names by later workers, which obscures the original English, scientific, and Nepali names given by Hodgson.

His least recognized achievement, and one that had it been more widely known would have probably established Hodgson's pre-eminence among Indian ornithologists, was his collection of water-colour paintings. Housed in the library of the Zoological Society of London, and bound in large folio volumes, eight in all, are 1125 sheets on birds, and 487 on mammals.[122,123] These are the original colour illustrations and pencil sketches that Hodgson commissioned from his Nepalese artists. Sadly, virtually nothing is known about these extremely talented men—even the name of one, Tursmoney Chitterhar, is recorded with some uncertainty. Raj Man Singh probably produced the majority of the paintings and it is likely that he was trained and employed by Hodgson for many years. After his eventual return to India in 1845, Hodgson must have been joined by these artists in his east-Himalayan retreat. A later set of paintings was executed, largely a reworking and improvement of those now kept by the Zoological Society of London, with some new additions including 30 sheets of amphibians, reptiles, and fish. These refined versions eventually went to the Zoological Library of the British Museum (Natural History) in London, but their original purpose was to illustrate a work on Himalayan birds and mammals that Hodgson had hoped to publish.

As mentioned previously, Hodgson had subscribed to, and probably had with him in Nepal, John Gould's *A Century of Birds from the Himalayan Mountains*.[15] This very early work on Indian avifauna was possibly the inspiration for some of Hodgson's collection of artwork, and a comparison of the two sets of paintings tends to reveal certain similarities. Gould, in an attempt to give drama to his paintings, occasionally contorted the posture of a bird far beyond any natural range of movements. In the very same families of bird where Gould employs this technique, Hodgson's illustrations bear a similar unnaturalness.[15] However, his artists' later work was freer and more natural and closer in style to contemporary bird art, with its concern for verisimilitude. Certainly for its time, Hodgson's collection of paintings was as good as, and more accurate than anything published previously, and he was at pains to point out the paintings' truth to life in his private letters and public announcements related to the proposed work.

Further examination of Hodgson's collection of paintings and those of Gould reveal the difference between paintings produced from dry skins and those done from live or freshly killed birds. This is particularly noticeable with regard to the soft parts—bills, legs, feet, and eyes. Presumably Mrs Gould, who actually did the paintings for the *Century*, when faced with the task of their preparation, either guessed the colour of such perishable features as the eyes, or simply reproduced the faded colours she found on the dry skins. Hodgson and his artists, with the advantage of sketches or drawings done from fresh specimens or live birds, would capture the features and colours truest to the living bird.

In March 1837 on behalf of his son, Brian Hodgson senior approached John Gould to enlist his support in the production of his son's book. Notorious as a calculating businessman, and described by Edward Lear (at one time one of his employees) as a 'harsh and violent man', Gould emerges from this exchange of letters as shrewd and self-seeking. He would only consider a work solely on birds, not on birds and mammals as Hodgson had originally intended. He refused to rework Raj Man Singh's paintings, 'perfectly convinced', with all the racism of his age, 'that no work executed from the drawings of Indian artists will sell'. He was to restart the drawings from scratch and he was to write the descriptions and measurements, leaving Hodgson to complete the text on behaviour and habitat. Birds duplicated in Gould's own *Century* were to be simply copied into the new larger series, and the whole project would be offered to the public as an extension of the former work. To be fair to Gould, he would take much of the financial responsibility, a burden his partner was keen to avoid, but there was no doubt that Hodgson was being asked to play a secondary role in his own project.[16]

In his reply, Hodgson's father pointed out the considerable disadvantages which such terms forced upon his son, and in the margin of Gould's letter repeatedly scribbled 'Gould and Co!' as a wry comment on the artist's motives. It seemed ridiculous that Gould should write the bulk of the text when Hodgson, who had already prepared so much information in his un-published notes and who alone had had firsthand experience of many of these birds in their natural state, should contribute only their behaviour and habitat. He also complained about Gould's intention that he should continue to receive Hodgson's paintings and specimens without being bound to start the work for another eighteen months. He could thereby continue to use and benefit from Hodgson's material without any liability.[61] Hodgson, in Nepal, wrote at the top of his copy: 'I am not likely ever to assent to that proposi-tion'.[124] He disapproved of the large format of Gould's paintings in his previous works and thought the text in both the *Century* and the *Birds of Europe* was 'miserably trivial'. He wished to produce a working text for the scientific naturalist, while Gould saw his market among 'particular chaps at home who have ample means to gratify a taste for the superb in works of art'. Between the scientific purist and the artist-entrepreneur the project foundered.[136]

Hodgson did not give up the struggle to see his book in print, however, for he was well convinced of its excellent potential. Advertisements appeared in Indian natural historical journals in 1835, 1836, 1837, 1842, 1843, and 1844. Large expensive works of this kind usually relied upon the goodwill of subscribers, people who promised to purchase the work once it was finished and whose names were circulated with any related advertisement. Naturally, the presence of such prestigious titles as the Governor-General's, Lord Auckland, were excellent auspices for its chances of success. Eventually Hodgson accumulated 380 names in India, and canvassed their opinion on what sort of book it should be: one modelled on Gould's *Century* or a more modest octavo or quarto format. By 1843 the *Journal of the Royal Asiatic Society* announced that 'in the present month the first division of the Zoology of Nipal' was about to come forth. A similar announcement appeared in the *Calcutta Journal of Natural History* the following year, but after that there was nothing and the book never materialized.

In the winter of 1843 Hodgson had been more or less dismissed from Nepal and had returned to England. He must certainly have looked for ways of publishing his plates, and as late as 1859 in a letter to a Mr Hawkins, he writes of 'some mature plan such as he (Gray) himself may approve to be submitted to the Trustees for the publication as soon as maybe of the drawings or a selection of them.' Yet nothing ever came of all his effort: the paintings still remain as large privately-bound volumes on the shelves of two London institutions, illustrations unseen and largely unknown by the general public.[119]

Not surprisingly, Hodgson was disillusioned by his lack of success. He was a proud and generous man who would never deign to point out the generosity of his giving, but was deeply offended when it went unacknowledged by others. In a letter to James Prinsep in March 1836 he wrote: 'By my soul it is a d——d bore to be compelled to fawn and entreat for subscribers as if one was the obliged instead of the obliging party, when one undertakes to labour with pains and cost for the love of science!'[40]

To Sir Alexander Johnstone, the vice-president of the Royal Asiatic Society, he vowed himself 'willing to be the drudge of science . . . through one of her true ministers and interpreters . . . But with any common editor I would not choose to cooperate. I dread a vulgar reputation in such matters.'[37] Hodgson had already expended a large sum of money on the paintings and the collection of skins, and had offered 3000 rupees to assist publication, but he was unwilling to venture further. Perhaps the memory of his father's fruitless speculation in Irish mining and his own considerable financial commitments to his family, both in England and India, held him back from Gould-like enterprise. It was probably for this reason that his book never saw the light of day, yet Hodgson, perhaps unfairly in this particular episode, saw himself as let down by his scientific brethren.

His disillusionment and sense of being cheated was a constant theme in his later correspondence with scientific individuals and institutions. He himself was of the highest integrity, and the treatment he received at the hands of others must have appeared decidedly shabby in the light of his own impeccable standards. The Grays, as we have seen, had meddled with the labels on his skins and had tampered with his own taxonomic division of the collection. Both the East India Company museum and the British Museum had very badly neglected his donations, while a number of the valuable paintings placed in the care of the Asiatic Society of Bengal and the Zoological Society of London had been temporarily lost. His main criticism, however, was levelled at those 'museum men' and 'closet naturalists' who attempted to reap for themselves whatever honours might have accrued from the discoveries of a field worker like Hodgson.

Foremost among these was Edward Blyth, the curator of the Asiatic Society of Bengal and a renowned ornithologist in his own right. In 1845, Hodgson complained to the Society that a paper he had submitted to their curator in May 1843 containing his descriptions of many new genera and species, had remained unpublished. Meanwhile, the contents of the papers were appearing as Blyth's own discoveries in the curator's report to the society.[7,8] Although this report bore the date December 1842 it was not, in fact, published until June 1844, presumably because Blyth, in an attempt to avoid suspicion, wished to suggest that 'his discoveries' predated those of Hodgson. At least nine species of bird new to science, including the exotic Fire-tailed Myzornis *Myzornis pyrrhoura*, whose discovery was credited to Blyth, were in fact 'poached' from Brian Hodgson.

Although this was by no means the only example of such malpractice, it is probably one of the worst that Hodgson encountered. The Society's reply to Hodgson's complaint, written by its secretary, Torrens, is interesting for the light it sheds on the business of poaching which a later ornithologist, Ticehurst, felt 'inseparable from scientific enterprise'. Torrens wrote: 'The man is cursed by that wretched professional English jealousy which insists on denying all merit and if possible interfering with all success of any other naturalists, more especially one not professional. The course of suspending him has been seriously discussed and I am ready to carry it through for the sake of annoyance I undergo from the jealousy, arrogance and absurd preten-sions of all my subordinates.'[146] Another man whom Hodgson accused of that 'wretched professional English jealousy' was John Gould himself.

It seems likely that Gould, who had dedicated his *Century* to the Zoological Society of London, enjoyed close links with it and had gained access to Hodgson's skins held in the Society's custody. Four new species of bird from Nepal, described by Gould between 1836 and 1838, were again almost certainly birds found by Hodgson, since he was the only person collecting in that country at the time. It was only when Hodgson returned to England in 1844 and again after his Darjeeling period in 1858 that he realized the full extent of others' malpractice. In 1859 he wrote to an associate complaining of 'that interloper and rogue Gould who has constituted himself the guardian of my stores in London'.[87]

In addition to Hodgson's personal grievances against individuals he also made constructive criticisms of the methods of natural scientists in general. It was customary during the nineteenth century for collectors in foreign countries to go on brief expeditions to a locality, shoot and skin as many specimens as they possibly could, then ship all those dead animals back to Europe where they would be worked upon by museum men. Hodgson argued that, based upon such insufficient evidence as the dried skins, the inferences and speculations of 'closet naturalists' would inevitably lead to a host of errors in taxonomy. He also called for a closer liaison between the societies and museums, and the naturalists on location. If learned societies truly wished to advance the course of knowledge then they would feel duty-bound to lend their assistance rather than attempt to anticipate the latter's discoveries. 'The first care of zoological societies should be to quicken and guide the attention of such local observers . . . and the last care should be unnecessary intrusion upon the province of these observers with no other object than to rob effectual enduring diligence of its reward.'[73]

In spite of a silver medal awarded to him by the Zoological Society of London in 1859 for his help in procuring a pheasant collection, on his return to England Hodgson's interest in natural history seems to have been dulled. Having made such extensive researches, one might have hoped that his homecoming would have spurred him on to the consummation of all those labours and the production of his 'Zoology of Nipal'. His final efforts, however, seem to have fizzled out and he was brought to the end of his tether. One more small insult, he claimed in a letter, and he would burn the lot! Fortunately, he did not lose sight of his own high ideals, and in 1870 when Alan Octavian Hume, the eminent Indian ornithologist, asked for assistance, Hodgson responded with characteristic open-handedness. He agreed to send all his unpublished portfolios, including the valuable paintings, back to India to Hume in Agra. Hume acknowledged the debt by dedicating the work, *The Nest and Eggs of Indian Birds*, to Hodgson, 'for the invaluable services which he rendered to Indian Ornithology'.[126]

Four years later the Zoological Society of London made Hodgson an Honorary Member, but after 1874 he ceased to play a significant role in the field of natural history. Perhaps he felt that he had played his part and now, quite simply, he wished to retire. It would be difficult to sum up briefly what had been Hodgson's importance in nineteenth-century ornithology; so many things that should have secured his reputation or signalled his triumph were taken from him both by the connivance of others and a tissue of adverse circumstances. Given this situation, it is perhaps best to let Hodgson speak for himself in an unusual moment of self-acclamation:

No-one knows better what pains and cost have been bestowed by me upon zoology and no-one better knows how little fruit of said pains and cost I have yet realised owing to the indifference of the so-called patrons of science . . . I must have stood without dispute the greatest discoverer nearly on record—certainly by far the greatest on record for northern India for my particular field.'[119]

4

A student of ornithology

HODGSON's ornithological work was undoubtedly his most important zoological achievement. (In his ornithological papers he provided numerous descriptions of birds including eighty of species new to science.) A complete list of these species he first described and of the other bird species he discovered is given in the Appendix. One significant aspect of his contribution was the emphasis he gave to observing birds, not only as dead specimens on a museum bench, but also as living creatures in their natural environment. He was interested, whenever possible, to record a bird's behaviour and habitat, and believed that these aspects of its biology were often clues to its taxonomic status, an important innovation amongst British ornithologists. Although Hodgson's papers consisted largely of detailed descriptions of both internal and external structure, in accordance with the traditions of his day, he also took notes on his field observations of wild birds. The following chapter is a distillation of his published ornithological material, paying particular attention to his work on behaviour and habitat. By drawing it together, hopefully, Hodgson's own ambitions are in a very small way fulfilled, since he had wished to complement the unpublished paintings with this sort of writing.

When one contrasts these pieces with modern standards of scientific writing, and indeed with his own style of prose in the other work, they stand out as a pleasant source of light relief. Take, for example, his writings on the Bengal Florican *Houbaropsis bengalensis* (Plate 6), a striking bustard of the Nepalese lowlands, and perhaps Hodgson's favourite bird, or his anthropomorphic portrayal of the Great Pied Hornbill *Buceros bicornis*. Then there are his descriptions of Long-legged Buzzard *Buteo rufinus*, and Black-shouldered Kite *Elanus caeruleus*, both delightful vignettes of hunting birds of prey.

Hodgson had a particular interest in raptors and was often critical of their classification by other ornithologists. He acquired a comprehensive collection, consisting of 53 of the 56 species now on the Nepalese list. Eighteen of these he described, including the Mountain Hawk-Eagle *Spizaetus nipalensis* which was new to science.[50] He collected several specimens of this species over a period of ten years and was understandably confused by its 'divergence of appearance with reference to sex, age and season'. This medium-sized forest eagle was described by Hodgson as 'adhering exclusively to the wilds and killing its own prey, which consists of pigeons and junglefowl and partridges'.[52] Of the Black Eagle *Ictinaetus malayensis*, he noted its very small outer toe and claw, believed to be adapted for nest robbing.[88] Recognizing their common adaptations to fishing, he wrongly placed both Pallas's Fish Eagle *Haliaeetus leucoryphus* and Grey-headed Fishing Eagle *Ichthyophaga ichthyaetus* in the same genus as Osprey *Pandion haliaetus*. 'Pandion is the King of fishers, and a more beautiful instance of the adaptation of structure to habits than this genus exhibits is not to be found in the whole circle of ornithology.'[68] Then comes his portrait of the dainty, grey and white, Black-shouldered Kite: 'The large eyes are suited for hunting. Its flight is agile and silent, aided by its long wings and soft plumage ... Commonly it is seen skimming the cultivation, occasionally poising itself on the wing for the purpose of getting a distincter view of some mouse, small bird, or insect which has stirred on its beat and upon which, when clearly perceived, it stoops perpendicularly with the speed of lightning.'[71] He watched the technique of the Long-legged Buzzard, saying 'after harvest it comes into the open country, and is perpetually seen in the fields perched on a clod and looking out for snakes which constitute its chief food. It also preys on rats and mice, and on quail, snipes and partridges; but is reduced to take the birds on the ground. I have seen it however, make a splendid stoop at a quail, which after being flushed, chanced to alight on a bare spot, so as to be visible to the bird as he followed it with his eye on the wing and marked it settle.'[50]

The only new gamebird Hodgson discovered in Nepal was the attractive Snow Partridge *Lerwa lerwa* (Plate 10). He 'found it close to the permanent snows, among rocks and low brushwood, where it sustains itself upon aromatic buds, leaves and small insects ... gregarious in coveys, nestles and breeds under jutting rocks'.[33] The Common Hill Partridge *Arborophila torqueola* was described as 'exclusively a forester, inhabiting the interior of deep woods. Gregarious in coveys, breeds on the earth, feeds on the ground and on trees eating berries, seeds and insects. Has a shrill twittering call. Is very timid and not at all pugnacious.'[72] By contrast, the Chukar Partridge *Alectoris chukar* was 'on elevated bare, dry, stony slopes; lays many eggs of a white colour, in a careless nest, after the manner of partridges ... the males are famous for courage and pugnacity.'[72]

The Bengal Florican was obviously very special to Hodgson. It was probably the only species for which he employed Nepalese observers to watch the birds closely. Prior to his paper almost nothing was known of the bird, and it still constitutes a large part of our knowledge of this now rare and threatened bustard. It is the most detailed account he wrote of any bird species although unfortunately some of the information, including a description of the male's courtship display, has later been found to be inaccurate. 'Of all Indian game birds it is the most striking to the eye and the most grateful to the palate.' The grasslands it favours are 'not so thick nor so high as to impede the movements or vision of a well-sized bird that is ever afoot and always sharply on the lookout. Four to eight are always found in the same vicinity . . . and the males are invariably and entirely apart from the females, after they have grown up. Even in the season of love the intercourse of the sexes among adults is quite transitory . . . when the rites of Hymen have been duly performed, the male retires to his company and the female to hers . . . nor is there any evidence the male ever lends his aid to the female in tasks of incubation and of rearing the young. The female sits on her eggs about a month, and the young can follow her very soon after they chip the egg. In a month they are able to fly, and they remain with the mother for nearly a year, or till the procreative impulse is again felt by her, when she drives off the long since full grown young. Two females commonly breed near each other, whether for company or mutual aid and help . . .'[104]

The remarkable wader, the Ibisbill *Ibidorhyncha struthersii* (Plate 8), was originally described in 1829 by Hodgson, who gave it the delightful name of the Red-billed Erolia. The only specimen he is known to have procured was shot on the banks of a stream in the Kathmandu Valley in October, apparently on passage.[38,39] Unfortunately, his paper was lost for a period, and the first description of the species was unfairly attributed to Vigors three years later. The Great Stone Plover *Esacus recurvirostris* (Plate 9), described by Cuvier, an eminent French professor of zoology, in the same year, was also a bird probably found by Hodgson. In his own paper he wrote on its 'strong, triturating gizzard, fitted with the aid of gravel to grind the crabs and other hard-shelled fish on which it feeds'.[58] Continuing with the waders, Hodgson gave accounts of all five Nepalese snipes, together with Eurasian Woodcock *Scolopax rusticola*. In his time they were mainly winter visitors and passage migrants to the Kathmandu Valley. He correctly guessed the latter was an altitudinal migrant, breeding in the northern Himalayas and wintering in the central and lower regions. He remarked on its boldness and diurnal feeding behaviour, quite different to that of European populations. Of the snipes, he was the first person to describe the Solitary Snipe *Gallinago solitaria* and the Wood Snipe *G. nemoricola* (Plate 7). The flight of the former he noted as 'exceedingly rapid and devious like that of the Common Snipe . . . his first flight often a very long one'. The Wood Snipe was called the Woodcock Snipe by Hodgson, as it resembled that species in its dumpy figure, compact short bill, and habits. A 'large, dark, wood haunting snipe . . . its flights are short and unwilling; and if alarmed it will quit its usual haunts upon the confines of woods . . . for adjacent thick cover'.[25]

Once again, the rewards of discovery eluded Hodgson when his first description of the Speckled Wood Pigeon *Columba hodgsonii*[51] was delayed and another naturalist received the credit, although its Latin name bears testimony to his involvement. Two other species of pigeon, both resident in the lower hills and terai, were the Mountain Imperial Pigeon *Ducula badia*, noted as 'exclusively arboreal and fruit-eating . . . almost a solitary', and the Thick-billed Green Pigeon *Treron curvirostra* (Plate 11): 'rarer and more shy than the other doves . . . adhering to the forests and feeding directly on soft fruits'.[44]

Hodgson wrote a paper on only one member of the cuckoo family, the Drongo-Cuckoo *Surniculus lugubris*: 'A very singular form combining all the essential internal and external characters of *Cuculus* (cuckoo) with the entire aspect of *Dicrurus* (drongo).'[77] However, he was particularly interested in owls, describing over half the species recorded in Nepal. These included for the first time the Oriental Scops Owl *Otus sunia*,[46] (Plate 12) the Forest Eagle Owl *Bubo nipalensis*,[46] (Plate 13) and the Tawny Fish Owl *Ketupa flavipes*.[56] He gave an account of the magnificent Forest Eagle Owl, the largest owl in the Himalayas, writing of its 'muscular power in the legs far exceeding that of the Eagles, and with talons capable of giving that power the utmost effect in the destruction of life. It tenants the interior of umbrageous woods, and by reason of the feeble light penetrating them even at noon-day, it is enabled to quest subdiurnally in such situations. It preys on pheasants, hares, rats, snakes, and sometimes on the fawns of the Ratura and Ghoral.'[46] The closely related Northern Eagle Owl *Bubo bubo* he called the Hole-haunting Owl, as its 'habitation is sometimes in a hole or burrow in a bankside . . .' It hunts 'commencing operations long after dark and carrying them on in open country'.[46]

The similarities in appearance of the Brown Hawk Owl *Ninox scutulata* (Plate 14) to hawks fascinated him: the 'decreased size of head . . . very small development of the ear conch and of the (facial) disc, the greatly superior development of the wings and tail and the greater firmness and still closer set of the whole plumage'.[70]

Presumably the difficulty of trapping or shooting such fast-flying aerial birds allowed Hodgson to collect only three species of swifts. These included the Little Swift *Apus affinis* which he described as the 'common swift of the central region, where it remains all the year, building under thatched roofs, and against the beams of flat roofs. It lays two white eggs and breeds repeatedly.'[60] It is interesting that he obtained the Dark-rumped or Khasi

Hills Swift *Apus acuticauda*, an enigmatic bird that has never been recorded in Nepal since, and is known only from the Khasi and Mizo Hills in north-east India. It is not mentioned in his papers nor listed by the Grays in their catalogues of his specimens, yet the species was first described from a Hodgson Nepal specimen by Jerdon.[131]

In his paper on the Blue-bearded Bee-eater *Nyctyornis athertoni* (Plate 15) Hodgson wrote 'These birds feed principally on bees and their congeners but they likewise consume great quantities of scaraboei and their like. They are of dull staid manners and never quit the deepest recesses of the forest. In the Raja's shooting excursions they are frequently taken alive by the clamorous multitude of sportsmen, some two or more of whom single out a bird and presently make him captive, disconcerted as he is by the noise.'[55] Equally original was Hodgson's correct surmisal that the hornbill genus *Buceros* was chiefly frugivorous and not meat-eating as previously thought. He wrote papers on the Rufous-necked Hornbill *Aceros nipalensis* (Plate 16) and Great Pied Hornbill *Buceros bicornis*, including the first description of the former. He gave an account of its nesting behaviour saying 'It inhabits decaying trees, the trunks of which it perforates from the side, making its abode within upon the solid wood, and having its mansion further secreted by an ingeniously contrived door.' He obtained two live birds, an adult and an immature, which were 'extracted from the tree by cutting down its nest with axes'.[22] His paper on the Great Pied Hornbill reads: 'This species is gregarious . . . of staid and serious manners and motion . . . perched on the top of some huge fantastic Bar tree, you shall see these large grotesque and solemn birds sit . . . for hours with the immovable gravity of judges, now and then exchanging a few syllables.'[31]

Hodgson was understandably impressed by the diversity of woodpeckers in Himalayan forests, collecting at least 20 species, showing 'every known modification of form'. In his papers he described six species, including for the first time the diminutive White-browed Piculet *Sasia ochracea*,[59] and the now uncommon Bay Woodpecker *Blythipicus pyrrhotis*.[67]

Among the passerines, he first described the attractive Silver-breasted Broadbill *Serilophus lunatus*[75] and the Blue-naped Pitta *Pitta nipalensis*.[66] Sadly, there was yet again a delay in the publication of Hodgson's papers resulting in the first description of the former species being attributed to Gould. He observed two hirundines, the Red-rumped Swallow *Hirundo daurica* and the Crag Martin *Ptyonoprogne rupestris*. He called the former 'the Common Swallow of the central region, a household creature remaining with us for seven or eight months of the year'.[60] On the other hand, the Crag Martin 'preferred rocky situations of the central and northern regions'.[60] Hodgson only wrote on one pipit, the Upland Pipit *Anthus sylvanus* (Plate

17), for which he provided the original description. He remarked on its behaviour: 'Exclusively monticolous; found in the brushy uplands of the central region; feeds and breeds on the ground; food grylli and other insects and seeds. Nest made loosely of grass and saucer-shaped; eggs bluish, thickly spotted.'[94] Another first for Hodgson was his original discovery of the Black-winged Cuckoo-Shrike *Coracina melaschistos*,[63] a summer visitor to the woods of the Kathmandu Valley. Three accentors, gems of the Himalayas and close relations of the European Dunnock *Prunella modularis*, are described in his papers, one for the first time—the Maroon-backed Accentor *P. immaculata*. Hodgson commented 'These birds are much on the ground and have an ambulatory structure of legs and feet. They are found in the central and northern regions.'[94]

Inevitably some birds presented an easier target to Hodgson's collectors than others. This was obviously the case with three conspicuous or noisy groups: thrushes, flycatchers, and babblers. Hodgson obtained an extensive collection of 53 species of the thrush family, eight of which he described for the first time: the Indian Blue Robin *Luscinia brunnea*,[66] Golden Bush Robin *Tarsiger chrysaeus*,[94] White-tailed Robin *Cinclidium leucurum*,[94] (Plate 18) Grandala *Grandala coelicolor*,[90] (Plate 19) Purple Cochoa *Cochoa purpurea*,[54] (Plate 20) Green Cochoa *C. viridis*,[54] (Plate 21) Black-backed Forktail *Enicurus immaculatus*, and Slaty-backed Forktail *E. schistaceus*.[49] He wrote of the beautiful Golden Bush Robin: 'dwells in low brushwood solitarily, feeds chiefly on small ground insects. Makes its nest on the ground, saucer-shape of moss, and places it under cover of some projecting root or stone; eggs verditer'.[94] The Grandala he considered to be 'A singular bird having the general structure of a thrush but with the wings vastly augmented in size and the bill of a Sylvie.' It inhabits 'the northern region in under spots near snow' and the male is noted for his striking purple-blue plumage.[90] Concerning the Purple and Green Cochoas he commented: 'shy in their manners, live solitarily or in pairs, breed and moult but once a year, nidificate on trees, and feed almost equally on the ground and on trees. I have taken from their stomachs several sorts of stony berries, small univalve molluscs and sundry kinds of aquatic insects'.[54] Hodgson was particularly conversant with another member of the thrush family, the Asian Magpie-Robin or Robin Dayal *Copsychus saularis* (Plate 22), even today a common bird in Kathmandu and one he must have observed in the residency grounds: 'they are nowhere so common as in gardens and on lawns, which they enliven in spring by their song, and at all times, by their vivacity and familiarity. The Dahils are perpetually in motion, and raise and depress the body with flirtation of the tail, exactly in the wagtail manner.'[49] His flycatcher collection from Nepal was even more complete, comprising almost all of the 30 recorded species, and for which he wrote first descriptions of five: the Rufous-bellied Niltava *Niltava sundara*,[65] Ferruginous Flycatcher *Muscicapa ferruginea*,[94] Slaty-blue Flycatcher *Ficedula tricolor*,[94] White-gorgetted Flycatcher *Ficedula monileger*,[94] and Orange-gorgetted Flycatcher *Ficedula strophiata*.[65]

However, perhaps Hodgson's most comprehensive collection, and thorough study of any Himalayan family, was that of babblers Timaliidae. This collection totalled 71 species for which he wrote original descriptions of 28,[42,45,48,57,66,73,76,83,90,93,94,99,147] while another ten species were first described by Blyth, the Grays, Gould, and Moore from Hodgson's specimens. In his paper on the large genus of babblers, the laughing-thrushes *Garrulax*, he gave considerable details on their behaviour: 'gregarious, noisy and alert. They frequent the deep and dank forests and groves exclusively: procure the greater part of their food on the ground, use the trees but for security when disturbed, for nidification, and for occasionally eking out their repasts with berries, pulpy fruit or caterpillars; and are for the most part, incapable of a sustained flight. Their habitat is very extensive since they are almost equally common in the southern, central and northern regions of these hills . . . Many of the species are caged and tamed with facility; and they are more often turned loose into walled gardens, whence they seldom attempt to escape, if there be a considerable number of trees, and where they can be of great service in destroying pupae, larvae and perfect insects, especially those which are generated or feed in manure.'[42] Hodgson also related the habits of two other genera of babblers, *Stachyris* and *Yuhina*. The latter, he wrote, are 'little birds preferring the lower and umbrageous trees, usually found in small flocks; and have a monotonous, feeble, monosyllabic note. Inhabit the central and northern regions . . . food viscid stony berries and small scaly insects such as harbour among foliage.'[45] The *Stachyris* babblers are 'arboreal . . . shy of man . . . feeding on tiny insects, larvae, and pupae. They build large globular nests, which are fixed upon and between the crossing twigs of thick low bushes, and lay four or five eggs of a pale fawn colour, either unmarked or spotted with brown'.[94] Parrotbills (Plate 23), those remarkable stubby-billed babblers and close relatives of that European misnomer the Bearded Tit *Panurus biarmicus*, were observed by him to 'inhabit the Kachar dwelling in thick brushwood in small flocks'. He collected all of the Himalayan species and wrote original descriptions of four.[73,83,90,93]

It comes as no surprise that the small fast-moving warblers, which present such serious identification problems to contemporary ornithologists, were an even greater trial for Hodgson. Lacking modern sophisticated optical equipment and faced with the difficulties of trapping or shooting such highly active birds, he was only able to observe four confiding or conspicuous species. In his notes on the aptly named Common Tailorbird *Orthotomus sutorius*, a familiar species of the Kathmandu Valley, Hodgson remarked on its strange nesting habits saying it 'makes a beautiful pensile nest, by sewing together the edges of large leaves. Dwells in low bushes and hedgerows and fences, picking up minute insects from the leaves and decayed wood, and frequently descends to the ground'.[94] Another conspicuous long-tailed warbler, the Striated Prinia *Prinia criniger*, a common resident of the middle and lower hills, was first named by Hodgson. Its favourite sites, he noted 'are those upland downs scattered with brushwood . . . it is almost perpetually on the ground seeking its favourite food . . . small scaled insects and their eggs'.[48] In his paper on the Nepalese tesias, he described the Grey-bellied Tesia *Tesia cyaniventer* and the Chestnut-headed Tesia *T. castaneo-coronata* and gave the first scientific name to the former. He wrote 'These little birds have a very strong muscular stomach and feed on hard grass seeds and hard minute insects . . . dwell in moist woods where there is plenty of underwood . . . they are solitary and live and breed on the ground.[66] In spite of all the problems which he faced with such a family, he still managed to collect 47 species from the Himalayas, including at least 36 from Nepal, rather more than half that country's total. He provided original descriptions for eight species,[48,66,94,147] of which, remarkably, four are skulking bush warblers that breed in the northern region: the Brown-flanked Bush Warbler *Cettia fortipes*,[94] Aberrant Bush Warbler *C. flavolivacea*,[147] Grey-sided Bush Warbler *C. brunnifrons*,[94] and Brown Bush Warbler *Bradypterus luteoventris*.[94]

Hodgson wrote several papers on specialist arboreal feeders, including first descriptions of three sunbirds and a spiderhunter.[147] The Sultan Tit *Melanochlora sultanea* 'a splendid bird, nearly twice the size of the Great Tit *Parus major* of Europe . . .' was another first for Hodgson. 'It explores foliage' he wrote, 'feeding upon soft arboreal insects . . . exceedingly fond of caterpillars.'[73]

The shrikes and crows are well represented in Hodgson's collection, totalling 30 species and including most of those on the Nepalese list. He accurately related the behaviour of shrikes: found 'in open country . . . have their perches on the upper and barer branches of trees and bushes, whence they descend to seize their prey on the ground . . . food all sorts of hard and soft, flying and creeping insects, and their larvae and pupae, also small lizards, feeble birds, mice and almost anything the birds can master. They have harsh voices very like the kestrel's . . . bold and daring in manner'.[64] The Crow-billed Drongo

Dicrurus annectans (Plate 24), which still poses a number of questions to present-day ornithologists, was first found by Hodgson although he admitted that he knew little of its habits and status.[63] In contrast, he wrote in detail of the Black Drongo *D. macrocercus*, a common bird throughout the subcontinent. 'It is familiar with man', he said, 'and seems to love the neighbourhood of country houses. Commonly makes short jerking parabolic flights from and to a bare tree, whereon it sits watching for insects and thence darts, as above described, to catch them on the wing. It is very bold, frequently pursuing Crows and Kite, that come near its perch—and such is the rapidity of its flight that it can overtake the Kite when he uses his best efforts to outfly it. It is very vivacious, darting about all day, and all night too, when the moon shines . . . it utters an agreeable whistling note of two prolonged syllables.'[28] Finally, in his work on Nepalese birds, we come to Hodgson's work on finches, of which he found and described eight species for the first time including the striking Scarlet Finch *Haematospiza sipahi* and Spot-winged Grosbeak *Mycerobas melanozanthos*[43] (Plate 25). The closely related munias were also observed in his papers: the Striated Munia *Lonchura striata*, Scaly-breasted Munia *L. punctulata*, and Chestnut Munia *L. malacca* (Plate 26). 'Many of them breed in the residency grounds and solitarily so far as I have observed. The nest is merely a large ball, laid against or upon naturally blended branches or stiff leaves, and having a small round entrance either on the side or at the top. The nest is composed of grass fibres, or leaves of the *Pinus longifolia* . . . The male and female labour at the work with equal assiduity, and share equally the task of rearing.'[43]

In addition to these papers, based almost entirely on Nepalese birds, Hodgson acquired a small number of species from Tibet, on Nepal's northern border. First of these was a magnificent male White Eared Pheasant *Crossoptilon crossoptilon* (Plate 27), perhaps the most striking of all that country's birds, and brought back to Kathmandu by a Nepalese envoy to Peking. Hodgson's description of this pheasant in 1838 was the first ever written for science on a Tibetan bird.[74] Lady Amherst's Pheasant *Chrysolophus amherstiae* and the Golden Pheasant *C. pictus*, both prized by aviculturists for their gorgeous plumage, were among his collection. However, the origin of both birds is uncertain, and since the latter does not occur in Tibet, it probably came from further east in China. In honour of his English wife, Hodgson gave the scientific name for another gamebird which he discovered, the Tibetan Partridge, *Perdix hodgsoniae* (Plate 28). His paper on this beautiful bird, the last of his ornithological studies, was published in 1856, and was based on a specimen presented to him by Nepal's chief minister, Jung Bahadur.[116] The four other Tibetan species in Hodgson's collection were the high-altitude gamebird Himalayan Snowcock *Tetraogallus himalayensis*, and three familiar crows of the Palearctic region, the Common Magpie *Pica pica*, Eurasian Nutcracker *Nucifraga caryocatactes* (Plate 29), and Common Raven *Corvus corax*.

5

Changes in birdlife since Hodgson's time

FOR the rest of the nineteenth century after Hodgson's departure, and indeed until midway through the twentieth century, Nepal remained a forbidden land to most Europeans. The notoriety of its malarial swamps, together with the Nepalese government's own effective restriction on the entrance of Europeans, sealed its borders almost completely. This in turn gave rise to a reputation for mystery; Nepal became a lost civilization. Those privileged enough to slip through the closed gate helped compound the myth, with their romantic tales of a fierce and colourful kingdom untainted by Western civilization, and of awesome mountain scenery surrounding the exotically templed city of Kathmandu.

It is not surprising that, when Nepal finally opened its doors to the modern world in 1953, this mythological status as a forbidden land played a significant part in the development of the country's tourist industry. The increased facility of air and road travel, the discovery that the Nepalese were some of the most hospitable people in Asia, and the astonishing beauty of the Himalayas, spread the reputation world-wide. For the ornithologist, in addition to all this, there is the country's rich avifauna. Today, Nepal has become a high priority destination for many European birdwatchers, and names such as 'Jomosom' and 'Langtang'—those of trekking routes through the Himalayas—have become familiar words to most people interested in Oriental birds. Of an equally high international reputation is the Royal Chitwan National Park, the country's first wildlife sanctuary. These localities form part of a well-trodden 'circuit' in India and Nepal, followed by birdwatchers from all over the world. With their expensive optical paraphernalia and a wealth of ornithological literature, these Western ornithologists have made some areas of Nepal the most intensively watched places in Asia.

Yet the recent rapid growth in popularity and understanding of Nepalese birds, although it builds upon and extends Brian Hodgson's early work, does not diminish his achievement. On the contrary, his enormous collection of 621 species, almost three-quarters of the country's present total, seems all the more remarkable when one considers his disadvantages. True, he was resident in the country for almost 23 years, yet he completely lacked the optical aids which are so indispensable to modern field studies. Unlike the contemporary birdwatcher, his travels in Nepal were seriously curtailed, not only by Court edict, but by the limited means of transport in those areas he could actually visit, and by the frailty of his own health. He had very little literature to help him identify and classify his specimens, but relied almost entirely on his own guns and trappers to secure them, and then his own judgement and investigations to make sense of what he observed.

At the same time, the addition of over 200 species to the Nepal list since Hodgson's time, and the contemporary growth of literature on Himalayan avifauna belies an unhealthy trend in the status of many of the birds themselves. As a result of the increase in population over the last century, the vegetation of much of central and eastern Nepal has since undergone large and accelerating changes. With the exception of parts of the lowlands, the west is less affected than elsewhere in the country and has a low human population. Much of this region is still extensively forested. However, in Hodgson's day most of the country up to a height of 4500 m was covered in forest, the dry northern slopes of the Himalayas on the Tibetan side probably being the only exception. Hodgson stated in 1856 that with a poulation of between three and five million, the vast majority of Nepalese land was unoccupied by man.

The following account of vegetational and demographic changes in Nepal has been presented in accordance with Hodgson's own division of the Himalayas into the lower, central, and upper regions.[114] Since Hodgson referred to the location of his birds almost entirely in terms of this system, some understanding of it is necessary to place in context the latter part of this chapter on the changing status and distribution of birds. The lower region he classified as the level of the plains up to 1220 m above the sea. This he redivided into the terai, the bhabar, and thirdly the range of sandstone hills known as the Siwaliks, with their intersecting ridges or duns. The central region comprised these mountain ranges between 1220 m and 3050 m above sea-level, while the upper region or Kachar was that between 3050 m and 4880 m.

Lying at a slightly lower altitude than the Gangetic plain, the terai, a narrow belt of malarial borderland 16 to 32 km wide, was described by Hodgson as a 'moist and rarely redeemed tract of level waste extending outside the sal forest along the base of the sub-Himalayas from the debouche of the Ganges to the Brahmaputra'. It has a tropical climate and was overgrown with long grasses in the rainy season, when floods frequently swept away existing roads. From early April to mid-October the terai was uninhabitable to all except the Awalia people, who took their name from the deadly 'awal', to which they were resistant. These people provided an invaluable service to Hodgson by carrying his letters to and from Kathmandu, so maintaining his links with the outside world. However, their monopoly on this largely fertile land was broken with the eradication of malaria in the 1950s, when the population increased dramatically. As a result of intensive cultivation the remaining grasslands now virtually all lie within national parks and reserves. Apart from Sukla Phanta in the far western lowlands which covers 390 hectares, only a small number of patches each of limited area still exist.

The area immediately north of the terai which rises up to an altitude of 300 m is known as the bhabar, and also has a tropical climate. However, it is as dry as the terai is wet, due to the porous nature of its gravelly soil, and supports a light, open forest with little undergrowth and few tree species, dominated by one, Sal *Shorea robusta*. According to Hodgson the Sal forest is 'entirely void of cultivation and is a prodigious assemblage of noble trees . . . with an average breadth of twelve miles'. Today, it still predominates over wide areas of the central and western bhabar.

Beyond the bhabar zone lie the first outer Himalayan foothills, called by Hodgson the Churighati range and now termed the Churia or Siwalik Hills, which run east–west and rise to about 1220 m. To the north, lies a series of longitudinal sub-Himalayan valleys separated by narrow ridges, called duns. Further north, in an area of overlap between Hodgson's lower and central regions, lies the Mahabharat range, rising to 2740 m. Last century, these two ranges and intervening valleys were covered in dense forests, subtropical at lower altitudes and temperate higher up. In the monsoon season the region was as malarious as the terai and bhabar, and inevitably had a low population. Now, however, this altitudinal zone has probably changed more than any-where else in the Himalayas and, with the exception of west Nepal, is now intensively cultivated and densely populated. Even on the very steepest slopes farmers have built terraces covering the hills from top to bottom, because of the present-day scarcity of land. Their construction is a remarkable achievement, but the narrower terraces are frequently washed away in the monsoon along with topsoil of overgrazed slopes. Consequently, large areas of central and eastern Nepal have suffered vast deforestation and serious soil destruction.

Hodgson's central region, between 1220 m and 3050 m, comprises a broad complex of hills and valleys including the Kathmandu Valley itself. The forests are both subtropical and temperate with species of oaks, chestnuts, rhododendrons, and Chir Pine commonly occurring. Hodgson wrote: it 'sus-tains the most superb trees and shrubs and grasses, in general too rankly luxuriant to afford wholesome pasture. Probably not a tenth of this region is under cultivation.' As in the lower hills, the central and eastern areas are now highly populated up to 2000 m, and permanent settlements are frequent up to about 2745 m. Most hillsides have been cleared for grazing or terraced for agriculture.

An exception to this is the Kathmandu Valley, which was richly cultivated long before Hodgson's time. It has been inhabited since at least the eighth or seventh century BC, when a fierce tribe, the Kirantis, invaded and established a kingdom in the Valley. Its fertility and the increased trade brought great prosperity, and Hodgson estimated the population at about 350 000 in his day. There were three cities on the Valley floor, Kathmandu, Patan, and Bhatgaon, each crowded with beautiful temples, palaces, and gardens. Legends tell us, and indeed geologists confirm, that it was once covered by a lake which formed particularly deep and rich soils in the Valley floor. Lying at an elevation of about 1350 m it enjoys relatively mild temperatures and a higher rainfall than the surrounding region. This combination of rich soil and equable climate resulted in it being particularly suited to cultivation, and most of the Valley was covered with crops almost throughout the year.

However, today it is more intensively cultivated and its appearance has greatly changed. In the nineteenth century the Valley floor was considerably wetter and supported many small but permanent marshes and numerous shallow streams. Now, as a result of improved drainage, the marshes no longer exist and the streams have decreased. Scattered over the central part of the Valley there was also once a number of small, densely-wooded knolls, the 'central woods' referred to in Hodgson's unpublished notes. Although some of these, such as Swayambhunath and Pashupatinath, still exist, their under-growth is either much reduced or has disappeared. The human population of the Valley has substantially increased, that of Kathmandu alone standing at well over half a million. This has resulted in extensive urban sprawl and the loss of many of its lovely gardens.

Hodgson stated that the 'mountains containing the Valley are covered everywhere with the noblest garniture of trees and copiously supplied with rills'. At the foot of the hills there was sloping, grassy ground with numerous areas of scrub jungle and small trees. Now forests on slopes encircling the Valley are much depleted, even in protected areas such as Nagarjung. Their undergrowth has been much diminished, and many trees have been felled. In general, remaining forests are now drier and more open, and the lower slopes are covered solely by scrub or secondary growth. The only forests retaining any luxuriant primary growth are on those slopes facing away from the Valley and parts of Phulchowki mountain south of Kathmandu.

Plate 1. Hodgson's Frogmouth, *Batrachostomus hodgsoni*.

Plate 2. White-bellied Redstart, *Hodgsonius phoenicuroides*.

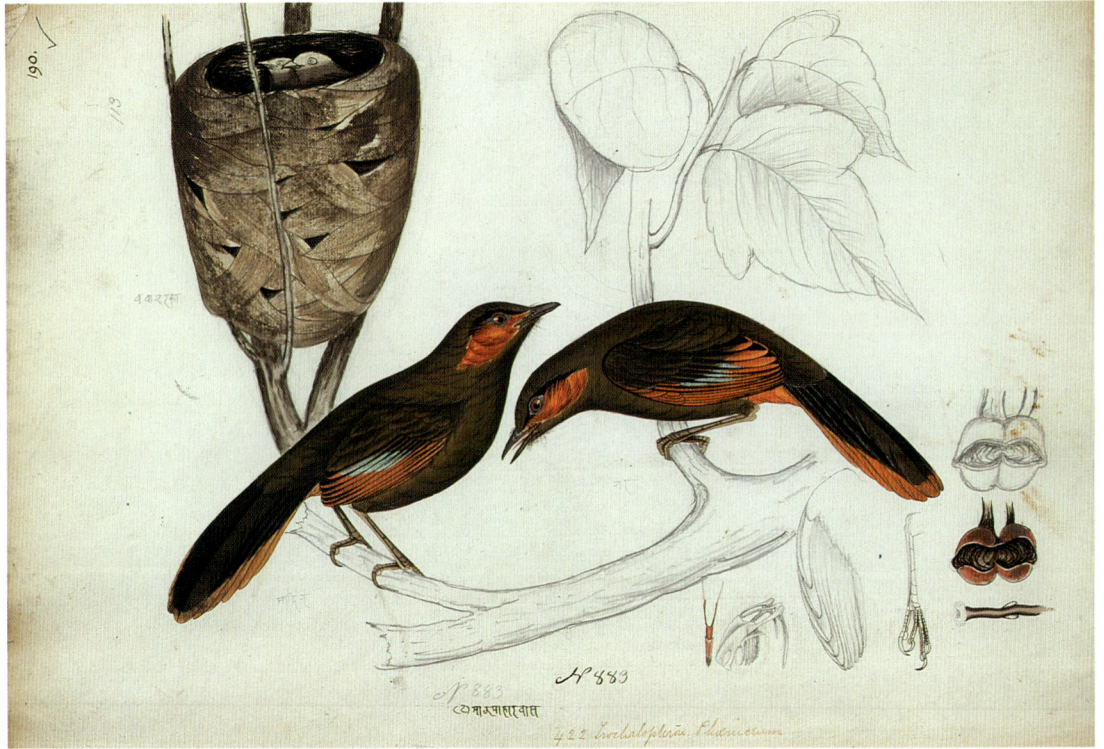

Plate 3. Red-faced Liocichla, *Liocichla phoenicea.*

Plate 4. Collared Treepie, *Dendrocitta frontalis.*

Plate 5. Long-billed Wren-Babbler, *Rimator malacoptilus.*

Plate 6. Bengal Florican, *Houbaropsis bengalensis.*

Plate 7. Wood Snipe, *Gallinago nemoricola*.

Plate 8. Ibisbill, *Ibidorhyncha struthersii*.

Plate 9. Great Stone Plover, *Esacus recurvirostris*.

Plate 10. Snow Partridge, *Lerwa lerwa*.

Plate 11. Thick-billed Green Pigeon, *Treron curvirostra.*

Plate 12. Oriental Scops Owl, *Otus sunia.*

Plate 13. Forest Eagle Owl, *Bubo nipalensis*.

Plate 14. Brown Hawk Owl, *Ninox scutulata*.

Plate 15. Blue-bearded Bee-eater, *Nyctyornis athertoni*.

52

600

4/8

(২) नास्त्राहाए हे औटा
(३) औशि यो नत्र गामैरह नाहिं चनैत्रा

146 Aceros nipalensis

Plate 16. Rufous-necked Hornbill, *Aceros nipalensis*.

Plate 17. Upland Pipit, *Anthus sylvanus*.

Plate 18. White-tailed Robin, *Cinclidium leucurum*.

Plate 19. Grandala, *Grandala coelicolor.*

Plate 20. Purple Cochoa, *Cochoa purpurea.*

Plate 21. Green Cochoa, *Cochoa viridis.*

Plate 22. Asian Magpie-Robin, *Copsychus saularis.*

Plate 23. Black-throated Parrotbill, *Paradoxornis nipalensis.*

Plate 24. Crow-billed Drongo, *Dicrurus annectans.*

Plate 25. Spot-winged Grosbeak, *Mycerobas melanozanthos.*

Plate 26. Chestnut Munia, *Lonchura malacca*.

Plate 27. Tibetan Eared Pheasant, *Crossoptilon crossoptilon.*

Plate 28. Tibetan Partridge, *Perdix hodgsoniae*.

Plate 29. Eurasian Nutcracker, *Nucifraga caryocatactes*.

Plate 30. Great White Egret, *Egretta alba*.

Plate 31. Lesser Whistling Duck, *Dendrocygna javanica*.

Plate 32. Purple Gallinule, *Porphyrio porphyrio*.

Plate 33. Mountain Hawk-Eagle, *Spizaetus nipalensis*.

Plate 34. Besra, *Accipiter virgatus.*

Plate 35. Bonelli's Eagle, *Hieraaetus fasciatus*.

Plate 36. Blue-breasted Quail, *Coturnix chinensis.*

Plate 37. Lesser Florican, *Sypheotides indica.*

Plate 38. Ring-necked Parakeet, *Psittacula krameri*.

Plate 39. Lesser Necklaced Laughing-thrush, *Garrulax monileger*.

Plate 40. Large Niltava, *Niltava grandis*.

Plate 41. Slaty-legged Crake, *Rallina eurizonoides.*

Plate 42. Pink-headed Duck, *Rhodonessa caryophyllacea.*

Plate 43. Laggar, *Falco jugger*.

Plate 44. Blue-naped Pitta, *Pitta nipalensis*.

Plate 45. Argali, *Ovis ammon.*

Plate 46. Musk Deer, *Moschus chrysogaster.*

Plate 47. Hispid Hare, *Caprolagus hispidus.*

Plate 48. Hodgson's Flying Squirrel, *Petaurista magnificus.*

Antelopidæ

द्वे-शेनियाबीलि

बेह॰चान॰पुलुक॰हेमल्का क्ष्योन्॰

Genus Pantholops nob
New species & type
Winter dress

Antelope Hodgsoni Abel
Pantholops Hodgsoni H.
Chuce of the Tibetans
Chiruc Unicorn of Europeans
Hab. Tibet

Plate 49. Chiru or Tibetan Antelope, *Pantholops hodgsoni.*

The upper limit of Hodgson's third region (4880 m) he estimated to be the average heights of the mountain passes and of perpetual snow. To him it was also 'nearly the limit of possible investigation', and he thought, 'the limit of the existence of organic phenomena', although plant life has been subsequently found as high as 5200 m. Unfortunately, Hodgson was unable to obtain personal experience of this region in Nepal, as it lies higher than the peaks of hills encircling the Valley to which he was limited. He referred to it as the Kachar or the snows, and described it as splendidly wooded with species including junipers, cedars, larches, dwarf rhododendrons, and birches. With the exception of dry valleys, heavy cloud cover prevents ripening of crops above about 2745 m, making permanent habitation which is dependent on agriculture impossible. Despite this, forests of the upper region are still depleted and degraded over large areas of central and eastern Nepal. The combination of summer grazing, collection of fodder for animals, fuel and building materials, often over long distances, has meant that probably few forests in these areas are now untouched by man.

In summary, since Hodgson's time there has been a sharp decrease in forest cover throughout much of Nepal. Despite recent afforestation programmes, less than one-third of the country remained under forest in 1980, according to government estimates, and forest losses are increasing.[3] Many remaining forests have been selectively thinned, and the understorey reduced by grazing livestock or the removal of foliage for animal fodder. In addition, undergrowth in some lowland forests is regularly burned, producing a fresh but reduced quantity of new growth for livestock.

The only area of Nepal which is documented ornithologically from Hodgson's time is the Kathmandu Valley and its encircling hills. Unfortunately, despite Hodgson's intimate knowledge of this region, he described the status of only a few species and never published a list of the birds he recorded there. However, his unpublished notes provide much information, and possibly he planned to incorporate this into his intended work on the zoology of Nepal. John Scully, who was resident surgeon between 1876 and 1877, was the first person to publish a list of birds occurring in the Valley, and their status.[141] His work was therefore a valuable complement to that of Hodgson's, and together they formed a fairly detailed picture of bird populations there during the last century.

The Valley was then an important staging post in the migration of wildfowl, waders, and storks, while some birds also remained to winter. As a result of drainage of the numerous marshes and streams, the variety and numbers of birds are now much reduced. In 1833, Hodgson wrote an interesting paper on the migration of what he termed Natatores (wildfowl, gulls, and terns), and Grallatores (cranes, herons, storks, waders, and rails), which he observed at Kathmandu.[29] This was the first account of trans-Himalayan migration and remained the only published work on migration in Nepal for over a hundred years. It is still a significant contribution to the knowledge of bird movements. He noted that these groups were generally only passage migrants on their way to and from the vast plains of India and Tibet, as the Valley was 'too small, dry, open and populous for their taste—especially that of the larger ones'. He described in detail months of arrival and departure: 'birds begin to arrive from the north towards the close of August. The first to appear are the Common Snipe (*Gallinago gallinago*), Jack Snipe (*Lymnocryptes minimus*) and *Rhynchoea* (Painted Snipe *Rostratula benghalensis*); next the Scolopaceous waders (except the Woodcock *Scolopax rusticola*); next the great birds of the heron, stork and crane families; then the Natatores; and lastly the Woodcock, which does not reach us until November. The time of re-appearance of these birds from the South is the beginning of March; and they go on arriving until the middle of May. The first which thus return to us are the snipes; then come the teal and ducks; then the Natatores; and lastly the great cranes and storks.' He noted the Grallatores were much more numerous than the Natatores, reflecting their greater abundance in the Indian plains. Most wildfowl only stayed short periods, a few days in spring and one or two weeks in autumn when the rice fields attracted them.

Unfortunately, Hodgson named birds almost entirely by genus only in his paper, but movements of many individual species can be deduced from his unpublished notes.[122] He recorded at least 15 species of these groups which have not been found there later. These included the Spot-billed Pelican *Pelecanus philippensis* and Greater Flamingo *Phoenicopterus ruber*, both transient visitors, the Glossy Ibis *Plegadis falcinellus* which stayed only a few days on passage, and the storks Greater Adjutant *Leptoptilos dubius* and Lesser Adjutant *L. javanicus*. The Great White Egret *Egretta alba* (Plate 30) and Intermediate Egret *E. intermedia*, both found by Hodgson,[122] have only been reported once this century.[129] The Black Stork *Ciconia nigra* was a common winter visitor and the Woolly-necked Stork *C. episcopus* common in summer according to Scully,[141] while both are only passage migrants to the Valley today.[129]

The wildfowl he collected included a single Whooper Swan *Cygnus cygnus* in January 1829, a vagrant winter visitor to the Indian subcontinent.[122] Another particularly interesting find was the now extinct Pink-headed Duck *Rhodonessa caryophyllacea*.[122] No records of either species in Nepal exist after Hodgson's day.[129] He found 14 other wildfowl species in the Valley, including many common European ducks. Most species moved through on passage but the Eurasian Wigeon *Anas penelope* and Common Teal *Anas crecca* remained to winter. Other interesting species include the trans-Himalayan migrants, the Bar-headed Goose *Anser indicus* and Greylag Goose *A. anser*; also the Lesser Whistling Duck *Dendrocygna javanica* (Plate 31) and the now scarce Common Shelduck *Tadorna tadorna*, neither of which have been found in the Valley subsequently. It is unfortunate that Hodgson gave no details of their abundance, but his duck paintings bear their date of collection. Most were taken in the three-week period between mid-October and early November, the notable exception being Garganey *Anas querquedula*, which passed through much earlier in August and September. All Hodgson's ducks and geese on their return passage north were obtained in March.[122] Scully provided a more detailed account in 1877, adding the Northern Pintail *A. acuta* which he described as the commonest duck, and also the Garganey, as wintering wildfowl in the Valley. He found three species additional to Hodgson's list of migrant ducks: the Ferruginous Duck *Aythya nyroca*, Ruddy Shelduck *Tadorna ferruginea*, and Northern Shoveler *A. clypeata*.[141] However, by 1948 as a result of extensive drainage, thousands of duck were noted flying over the Valley on migration, since there were few suitable places for them to alight.[137]

In his addition to his work on wildfowl, Hodgson recorded no less than 28 wader species in the Kathmandu Valley, five of which have not been found there since, including the Pied Avocet *Recurvirostra avosetta* and Whimbrel *Numenius phaeopus*. Indeed, three species, the Grey Plover *Pluvialis squatarola*, Greater Sand Plover *Charadrius leschenaultii*, and Eurasian Oystercatcher *Haematopus ostralegus*, have not been recorded in the country subsequently.[122, 129] Most of these waders were passage birds, mainly moving through in autumn. Other notable records include a Little Pratincole *Glareola lactea* in June, a local migrant usually found at much lower altitudes, and the Oriental Pratincole *G. maldivarum*, a scarce migrant which he found in the same month.[122] Another species not found again in the Valley, nor indeed in Nepal until 1981, is the Lesser Sand Plover *Charadrius mongolus*.[129] The later records indicate that it is probably a rare spring migrant, stopping only for short periods.

No doubt Hodgson and Scully were pleased to find that the Jack Snipe, Common Snipe, and Pintail Snipe *Gallinago stenura* were all common winter visitors and particularly numerous on migration, for these fast-flying waders proved excellent sport for the British in Kathmandu. Indeed, some Common Snipe remained all year and even the Solitary Snipe *G. solitaria* was not uncommon in winter.[25] By 1947, however, numbers of the former three species were much reduced[139] and recent reports of Solitary Snipe are few.[129]

The Wood Snipe *Gallinago nemoricola* is unusual in that it seems to have decreased during the thirty years between Hodgson and Scully's time. Although Hodgson found it not uncommon in winter,[25] Scully noticed it only twice.[141] The only other record from the Valley was in 1950,[139] and it is now rare throughout Nepal.[129]

There is evidence that the now-scarce Black-tailed Godwit *Limosa limosa* and Spotted Redshank *Tringa erythropus* have decreased in the Valley, and, in fact, numbers of all wader species have probably declined because of habitat losses, although little information is available.[122,129] Only four new wader species have been found in the Valley since Hodgson's years of residency. Of particular interest is the Grey-headed Plover *Hoplopterus cinereus*, first recorded in the Valley in 1937,[4] and a regular and quite common winter visitor since at least 1961.[129,138]

A single species of gull and only one tern, River Tern *Sterna aurantia*, were noted by Hodgson as transient visitors,[29] both families today remaining rare passage migrants in the Kathmandu Valley [129] and elsewhere in the Himalayas.

Amongst the crakes which Hodgson found was a single Water Rail *Rallus aquaticus*,[122] now a vagrant to Nepal with only one other known record from the Valley.[129] Other crakes have much declined in the Valley, including Baillon's Crake *Porzana pusilla*, the Common Moorhen *Gallinula chloropus* and the Purple Gallinule *Porphyrio porphyrio* (Plate 32). The former, now merely a scarce passage migrant, was found by Hodgson in May and October,[122] while Scully described it as common from July to December.[141] The other two birds have only been reported there once since last century, obviously deterred like so many of the birds mentioned by the disappearance of their marshland habitat.[129] The Ruddy-breasted Crake *Porzana fusca*, although still nesting in rice fields as Hodgson described,[122] has suffered a similar decline and is now uncommon.[129]

Hodgson was probably the first person to record cranes migrating over the Himalayas. The Common Crane *Grus grus* was noted by Hodgson,[122] but there is only one later record from the Valley.[129] Unlike now, Demoiselle Cranes *Anthropoides virgo* sometimes stopped for a few days in spring and autumn,[122] although they still regularly pass over the Valley, often in flocks of several hundred birds, in the present day.[129]

By contrast with these largely aquatic families, all but one of the 38 raptor species recorded in the last century have been found subsequently, and an additional eight migrant species have been reported.[129] However, raptor numbers have probably also suffered decline due to reduced opportunities for scavenging and widespread deforestation, although definite evidence is only available for a small number of species. All seven of the vultures noted by Hodgson still occur, but in lower numbers.[129] The Red-headed Vulture *Sarcogyps calvus* has been most seriously affected; formerly a common resident it is now only occasionally seen on the rim of the Valley.[129] A small migrant population of Pallas's Fishing Eagles *Haliaeetus leucoryphus* occurred last century,[141] but there are no subsequent reports from the Valley.[129] Another species which was probably more frequent during the Hodgson period was the Saker *Falco cherrug*, there being only one other Valley record.[132] The pretty Red-necked Falcon *F. chicquera* was very common in the Valley and was reported by Scully as breeding in the Residency grounds.[141] In 1947 it was still found to be not uncommon.[5] Formerly it frequented groves and gardens or large, solitary pipal trees, and must have been adversely affected by habitat loss as it has only been found on three later occasions in the Valley, and is now generally scarce in Nepal.[129] It is astonishing that Hodgson found no less than 11 raptor species in the residency garden.[122] These included forest birds such as the Mountain Hawk-Eagle *Spizaetus nipalensis* (Plate 33) and Besra *Accipiter virgatus* (Plate 34), which now only occur on hills surrounding the Valley.[129] Scully added Bonelli's Eagle *Hieraaetus fasciatus* (Plate 35) and Laggar *Falco jugger*.[141]

As a result of intensive cultivation and increased hunting pressure, several small gamebirds have disappeared from the Valley this century. Scully noted the Chukar Partridge *Alectoris chukar* as common and breeding on grassy slopes,[141] but there is only one report this century.[129] The Common Quail *Coturnix coturnix* was found by Hodgson and Scully in great numbers from mid-October to mid-December and from late March to the end of April, synchronous with the ripening of the great spring and autumn crops. According to Hodgson, 'they abound in the ripe crops; and in hills they may always be found except during the rains'.[62] The species has only been reported once since then in the Valley, and is scarce elsewhere in Nepal.[129] At least six specimens of Black-breasted Quail *Coturnix coromandelica* were obtained by Hodgson in April, May, and August, including a female containing eggs. He described it as 'exceeding rare' but 'as regular in its comings and goings as the common species'.[62, 122] There are just two subsequent records from the Valley and none from the rest of Nepal.[129] Hodgson also found five tiny Blue-breasted Quails *Coturnix chinensis* (Plate 36), in April and May, including two females with eggs ready to be laid.[122] No less than ten specimens of the Yellow-legged Buttonquail *Turnix tanki* were taken by Hodgson. He noted it to breed in corn, the young hatching in late May and early June.[62, 122] Both of the latter two species have only been reported once in the Valley subsequently, and the Blue-breasted Quail is very rare in the country generally.[129] The Barred

Buttonquail *T. suscitator* which also occurred in the past, Hodgson described as shy and 'never entering cultivation in the hills'.[62] It has not been found in the Valley since last century.[129] The spectacular small bustard, the Lesser Florican *Sypheotides indica* (Plate 37), was procured by Hodgson on at least five dates between April and June.[122] It has probably also disappeared from the Valley, as there is only one later record.[13]

Two inhabitants of the foothills, the Silver-eared Mesia *Leiothrix argentauris* and Sultan Tit *Melanochlora sultanea*, were plentiful in the central dun in 1877 according to Scully.[141] Reports of the Mesia since 1970 are few, and there are only a handful of records of the Sultan Tit this century.[129]

Two unusual population changes are presented by the parrot family. The Moustached Parakeet *Psittacula alexandri* found by Hodgson and Scully from August to November,[141] perhaps wandering to the Valley in search of fruiting trees, is now confined to much lower altitudes.[129] In contrast, the Ring-necked Parakeet *P. krameri* (Plate 38) which did not occur in the Valley in Hodgson's time, has been increasing there since 1979.[129] Escaped cage birds may well have been the origin of this isolated population. As equally puzzling as that of the last family is the status of certain owls. Hodgson provides the only record of the attractive Oriental Bay Owl *Phodilus badius* for Nepal. However, he obtained the skin from a Kathmandu shop, so it is possible the bird may not have originated in the country.[122] The Short-eared Owl *Asio flammeus*, described as the common field owl of the Valley by Hodgson,[122] is now rare in Nepal and only one later report from the Valley is known.[129] Conversely, the Spotted Little Owl *Athene brama* must have colonized the Valley since Hodgson's day. Now its harsh chatter is commonly heard at night in Kathmandu, but last century its distribution was limited to the lower hills and terai.[129]

It is also strange that neither Hodgson nor Scully found the Chestnut-headed Bee-eater *Merops leschenaulti* or Large Pied Wagtail *Motacilla maderaspatensis* there, as both are regular summer visitors now.[129] Perhaps these open-country birds, in common with the Egyptian Vulture *Neophron percnopterus* and Spotted Little Owl, were unable to reach the Valley because of the extensive barrier of forests, while subsequent deforestation has possibly allowed a slow advance and colonization of these species. Other mysterious declines are of the Chestnut Munia *Lonchura malacca*, Striated Munia *L. striata*, and Scaly-breasted Munia *L. punctulata*, previously common monsoon visitors to the Valley.[122] There are few recent records of the former, and the other two are now seen much less frequently.[129] Their decrease is undocumented, but may well have been caused by the cage-bird trade. Loss of the beautiful Asian Paradise Flycatcher *Terpsiphone paradisi* as a breeding bird is particularly sad. The black and white male has a ridiculously long tail which waves behind him as he flies. Last century, this flycatcher was very common from April to September,[122,141] but there are few recent Valley reports,[129] its decline probably being due to the reduction of gardens and hedges.

There is little direct evidence of changes in forest bird populations in the Valley and on surrounding hills, because their status was poorly recorded last century. Only three species have apparently disappeared: the Purple Cochoa *Cochoa purpurea*, and both the Lesser Necklaced Laughing-thrush *Garrulax monileger* (Plate 39) and Greater Necklaced Laughing-thrush *G. Pectoralis* which have not been reported since the 1940s.[129] Although the area still has a rich bird fauna, forest losses must have caused an overall decline in bird numbers. Selective felling and clearance of the understorey from woods and forests must also have had a detrimental effect. Babblers, some warblers, and thrushes, which require dense undergrowth and moist conditions, must have been particularly affected. Almost no babblers now occur on the Valley floor.[129] Hodgson noted several species of laughing-thrushes in the residency garden[42] but they are now restricted to the encircling hills of the Valley. The removal of mature trees must have been unfavourable to woodpeckers.

However, changes have been beneficial to a few species. The more open woods are well suited to flycatchers, and the more extensive scrub growth on hillsides must favour birds such as the Striated Prinia *Prinia criniger*, White-cheeked Bulbul *Pycnonotus leucogenys*, and Grey Bushchat *Saxicola ferrea*.

The subtropical forests of Phulchowki mountain on the southern perimeter of the Valley are amongst the richest of their type in the country. Several very local or uncommon species occur there, such as the Grey-chinned Minivet *Pericrocotus solaris*, Large Niltava *Niltava grandis* (Plate 40), and the Blue-winged Laughing-thrush *Garrulax squamatus*. Although Phulchowki still holds the most diverse fauna and flora in the area, its forests are sadly becoming depleted at an accelerating rate.

Due to government restrictions on Hodgson's movements, evidence of bird population changes other than in the Kathmandu Valley are mainly circum-stantial. Nevertheless, it can be surmised that since about 65 per cent of Nepal's breeding birds utilize forests,[129] the continuing extensive deforest-ation must be causing an overall decline in bird populations. Species depen-dent on dense undergrowth, semi-evergreen or evergreen forests, mature or decaying trees, and those occurring at the edges of their ranges, have been most affected. In addition, widespread drainage of marshes and pools must also have been detrimental. While hunting may have caused declines in some gamebirds, this is probably less important than habitat losses.

It seems likely that changes in the central and upper regions have been similar to those which have taken place on the perimeter hills of the Valley, although in the latter region they have probably been less, as it has been less affected by man. Inevitably, as he was unable to visit the upper region or Kachar himself, and his trappers mainly worked below 3050 m, Hodgson only collected a relatively small number of birds from the area. He acquired 80 species, all of which still occur, including the spectacular Fire-tailed Myzornis *Myzornis pyrrhoura*.[122] The limited data in Hodgson's papers and unpublished notes are all that are available for the region last century.

Since there was such a high risk of malaria, the terai and bhabar were much less frequently visited by Hodgson's trappers, and his collections from these zones were correspondingly limited, although they are much richer in species than those from higher altitudes. In his unpublished notes, Hodgson mentions only about 100 species,[122] but over 580 are known in modern times from these areas.[129] A large proportion (40 per cent) of the birds he recorded were wildfowl, waders, and storks, reflecting the wetter conditions of the terai in his time. He noted the Slaty-legged Crake *Rallina eurizonoides* (Plate 41) and Glossy Ibis *Plegadis falcinellus*[122] which have only been recorded once subsequently.[129] He described the Pink-headed Duck *Rhodonessa caryophyllacea* (Plate 42), which he also recorded in the terai, as 'shy, resides in remote large Jhils, feeds at night'.[122] It seems reasonable to assume that water-bird numbers have declined throughout most of the terai, and that their distribution has become localized due to swamp drainage and the encroachment of agriculture. A notable exception to this is the development of extensive marshes, sand banks, and open water in the south-eastern terai of Nepal at Kosi Barrage. Since the barrage was completed in 1964 it has become by far the most significant wetland in the country, and is of international importance. Almost all Nepalese water bird species occur there, and a peak of over 50 000 ducks in mid-February has been estimated.[133, 135]

Turning to other large non-passerines, Hodgson acquired at least three skins of the Laggar *Falco jugger* (Plate 43) and found it breeding.[122] This falcon is now scarce and has probably declined in both Nepal and India. Scully observed that the Common Crane *Grus grus* and Demoiselle Crane *Anthropoides virgo* were frequent in the central lowlands in winter.[141] Although the latter is still common on passage, both are now scarce winter visitors to Nepal. The Sarus Crane *Grus antigone* was also common in the central lowlands[141] but is now found only occasionally in the west.[129] In Hodgson's time, a characteristic bird of the terai was the Bengal Florican *Houbaropsis bengalensis*, the terai being described as its 'favourite and almost exclusive habitat'.[104] Now it is restricted to the few remaining disjunct, grassland areas, where it is uncommon.

Bird numbers and species diversity of the duns and lower hills have undoubtedly been affected as a result of the severe deforestation which has taken place there since last century. However, information is available on changes in status of very few species. In his unpublished notes, Hodgson recorded over 190 species from the area,[122] including six which have never been reported since and are almost certainly extinct in the country.[129] These included at least two specimens of the Imperial Heron *Ardea imperialis*, one of which was from Hitauda. This heron frequents rivers running through forest and was described by Hodgson as shy and rare.[122] He also reported the Rufous-necked Hornbill *Aceros nipalensis* as very rare and shy and 'tenanting the deep woods covering the hills which overhang the great Saul forest. Its more peculiar haunts are the largest trees, especially such as are decaying.'[122] Three of the other species, Hodgson's Hawk-Cuckoo *Hierococcyx fugax*, the Silver-breasted Broadbill *Serilophus lunatus*, and the Green Cochoa *Cochoa viridis*, favour semi-evergreen or evergreen forests. The Jungle Bush Quail *Perdicula asiatica*, which he described as a rare migrant in the duns,[62] is now extinct in Nepal. Another species which has apparently disappeared since last century is the Long-tailed Sibia *Heterophasia picaoides*. Hodgson procured at least four specimens from the terai and lower hills,[122], while Scully described it in 1877 as 'tolerable common' at Nimboator, which lay in the lower hills south of Kathmandu.[141] With the exception of the Jungle Bush Quail, all of the above species reached the western limit of their ranges in Nepal.

There is evidence that at least ten other species acquired in the region by Hodgson have since declined, and are now locally distributed and scarce or uncommon. All frequent moist, dense forests often with thick undergrowth, and their decrease can be attributed to loss of such forests. Hodgson obtained at least seven skins of the Tawny Fish Owl *Ketupa flavipes*, a bird favouring forested ravines,[122] but there are only three subsequent reports from the country.[129] The Red-headed Trogon *Harpactes erythrocephalus* has probably only decreased recently. Hodgson collected at least fourteen specimens of this species, including one in breeding condition from Hitauda.[122] It was not uncommon in this area as late as 1970,[130] but has subsequently only been reported from three other localities including Chitwan where it is very uncommon.[129]

The magnificent Oriental Pied Hornbill *Anthracoceros albirostris* and the Great Pied Hornbill *Buceros bicornis* have been particularly affected, since they require extensive areas of suitable forest. Hodgson described the latter as occurring in the terai and lower hills from the Rapti valley in central Nepal east to Assam, and also penetrating into the mountainous interior following the courses of larger rivers.[31] He commonly found 20 or 30 together, but now the species is locally distributed in lowland Nepal and uncommon where it occurs.[129] Hodgson procured three specimens of the Long-tailed Broadbill *Psarisomus dalhousiae* in May and provided the only evidence of the species breeding in Nepal.[122] It was described as common in the central duns in 1947,[6] but there have only been about nine reports in the last ten years.[129]

Another species of bird of this moist, dense habitat which has decreased in numbers is the Blue-naped Pitta *Pitta nipalensis* (Plate 44). Hodgson acquired five skins from the lower hills and Kathmandu Valley,[122] but there are only a few other records, all from the Valley.[129] Equally the White-throated Bulbul *Criniger flaveolus* was taken on at least six dates by Hodgson from the lower hills and terai,[122] while Scully described it as common in the central dun in 1877.[141] Now, however, it is only seen there occasionally and is locally distributed in Nepal.[129] The Purple Cochoa *Cochoa purpurea* has shared a similar fate and is now restricted to the Mai valley in the wet eastern forests of the country.[129]

Further species which he collected in damp, thick forests of the terai and lower hills were the Thick-billed Green Pigeon *Treron curvirostra*, Mountain Imperial Pigeon *Ducula badia*, Vernal Hanging Parrot *Loriculus vernalis*, and Ruddy Kingfisher *Halcyon coromanda*.[122] All are now scarce in Nepal and it is likely that they have suffered similar declines to the previously mentioned species.[129]

Depletion of forests is not only causing a reduction of Nepal's diverse bird and other wildlife, it is also seriously damaging the lives of its people. In 1981 the human population was estimated at 15 million and growing at the alarming rate of over 2.5 per cent per annum.[3] Over 87 per cent of the people rely on forest resources and many are experiencing acute shortages of wood for construction and fuel, and of fodder for their animals.[3] Tourists significantly worsen the problem in some areas by unnecessarily burning large quantities of wood. Each year the monsoon ravages deforested slopes, washing away millions of tons of topsoil, and causing flooding with heavy losses of life and property in the lowlands.

To ensure the future of Nepalese people and their wildlife, conservation of remaining forests and widespread reafforestation of barren slopes is urgently needed. The active involvement of local communities will be essential for success. Tourists should also be obliged to use other sources of fuel than wood.

Although a relatively high proportion of Nepal's land area is especially protected for wildlife, it is vitally important that more national parks and reserves are created to maintain Nepal's exceptionally rich diversity of species. Nepal now has a National Conservation Strategy developed by His Majesty's Government of Nepal in conjunction with the International Union for Conservation of Nature and Natural Resources. The strategy emphasizes the need for the rational use of resources and aims to strike a balance between the needs of the growing population and those of nature conservation.

6

Hodgson's mammal papers

WHILE the thoroughness of his work on Himalayan birds and the comprehensive nature of his bird collections tend to overshadow all Hodgson's other natural historical achievements, he did not think of himself primarily as an ornithologist. Hodgson made an important contribution to the study of mammals. As well as giving meticulous accounts of their external and internal structure he also made detailed investigations of their behaviour, regarding these as clues to their taxonomy. Working at a time before Darwin, Hodgson did not enjoy the guidance provided by the former's theory of evolution. In Hodgson's day, relations between species were poorly understood and his ethological work was a novel approach to this problem.

Hodgson kept a number of mammals in the residency gardens in Kathmandu, such as Nepal's three species of monkey, and many of his behavioural observations were made on these captive animals which were probably retained solely for this purpose. However, from Hodgson's papers it is clear both that he also had firsthand experience of wild animals in their natural state and that he delighted in these encounters. Particularly colourful were his descriptions of the behaviour of the Lesser Panda *Ailurus fulgens*, Common Ghoral *Nemorhaedus goral*, and Pygmy Hog *Sus salvanius*, all of which he must have studied for some time. The following is a summary of his behavioural studies of Nepalese mammals as described in his papers.

Hodgson was particularly interested in carnivores, acquiring 40 species and providing first descriptions of the Yellow-bellied Weasel *Mustela kathiah*, Spotted Linsang *Prionodon pardicolor*, Small Indian Mongoose *Herpestes auropunctatus*, and Crab-eating Mongoose *H. urva*. His skin collection included all the Nepalese species, although he procured eleven of them from India and Tibet rather than Nepal.

In his paper on the Dhole *Cuon alpinus* he wrote: 'Of all the wild animals that I know of similar size and habits, the Buansu, which is large, gregarious and noisy in its hunting, is the most difficult to be met with. This prototype of the most familiar of all quadrupeds with man is, in the perfectly wild state, the most shy of his society. The Wild Dog preys . . . chiefly by day . . . six, eight or ten unite to hunt down their victim, maintaining the chase by their powers of smell rather than by the eye.'[32]

Little has been added to our knowledge of the Spotted Linsang or Tiger-Civet since Hodgson's study of this beautiful animal. It was 'very numerous in Nepal and Sikkim. Equally at home in trees or on the ground, it dwells and breeds in the hollows of trees. It is not gregarious at all, and preys chiefly upon small birds which it is wont to pounce upon from the cover of grass.'[108] He also described the habits of the closely related Common Palm Civet *Paradoxurus hermaphroditus* and Masked Palm Civet *Paguma larvata*.[41] The former is 'no more shy of inhabited and cultivated tracts than the Common Mongoose and its favourite resorts are the old and abandoned mango groves. In holes of the decayed trunks of the trees it seeks a place of refuge. However rapacious its ordinary habits—and those of few of the carnivora are more so—it feeds freely upon the ripe mango in season, as well as upon other ripe fruits; but its more usual food consists of live birds and of the lesser mammals, the former of which it seizes upon the tree as well as on ground with a more than feline dexterity.' The Masked Palm Civet frequents forests and was very common in the central region of Nepal. Its behaviour is similar to that of the last species. Hodgson kept one for a few years: 'the caged animal was fed on boiled rice and fruits which it preferred to animal food not of its own killing . . . when sparrows, as frequently happened, ventured into its cage to steal the boiled rice, it would feign sleep, retire into a corner, and dart on them with unerring aim.' Another member of the same family, the Large Indian Civet *Viverra zibetha*, was noted as dwelling 'in forests or detached woods and copses . . . a solitary and single wanderer . . . feeds promiscuously upon small mammals, birds, eggs, snakes, frogs, insects, besides some fruits and roots.'[84]

Almost all of our present knowledge of the behaviour of the Lesser Panda was provided by Hodgson. They are, he related: 'quiet, inoffensive animals, their manners staid and tranquil; their movements slow and deliberate. As climbers no quadrupeds can surpass, and very few equal them, but on the ground they move awkwardly as well as slowly. They are monogamous, and live in pairs or small families . . . feed on fruits, tuberous roots, thick sprouts such as those of the Chinese bamboo, acorns, beech mast and eggs, the last they are very fond of. But they love milk and ghee, and constantly make their way furtively into remote cowherds' cottages to possess themselves of these luxuries. They sleep a deal in the day and dislike strong lights.'[107]

Hodgson made valuable additions to our knowledge of the Bovidae, collecting almost all the Nepalese species and writing numerous papers on this family, which includes antelope, cattle, goats, and sheep. Of the Four-horned Antelope or Chouka *Tetracerus quadricornis* he wrote: 'Exclusively confined to primitive forests and to the parts where thick undergrowth abounds . . .

dwells in the forests at the base of the mountains. Found in pairs or solitarily. Monogamous. Very shy and when hunted bounds like the common antelope thence one of its names, from Chouk, a leap.'[109] The Gaur *Bos gaurus*, a magnificent animal related to domestic cattle, was considered 'never to quit the deepest recesses of the Sal forest. It is gregarious in herds of from 10 to 30 . . . there are usually two or three grown males whose office it is to guide and guard the party . . . they manifest a degree of shyness unparalleled among the bovines.'[69]

The three goat antelopes of the Himalayas described by Hodgson included the Common Ghoral: 'a stocky goat-like animal found all over the hills wherever there are precipices and crags, also in dells where rough and rocky . . . most wary and the least noise sends him over ground that makes your blood run cold.' The closely related Mainland Serow *Capricornis sumatra-ensis* was: 'a large, coarse heavy animal with bristly thin set hair . . . the body is short and thick, the chest deep, the head coarse and spiritless. It is seldom found in herds; and the grown male usually lives entirely alone, except during the breeding season . . . it tenants the central region. It rushes with fearful precipitancy down the mountains it inhabits.'[24,34] The third of this group, the Takin *Budorcas taxicolor* of which Hodgson was especially proud, was one of the last animals he discovered: 'A large, massive and remarkable animal, with Bovine proportions, very gregarious.'[112] It inhabits forests of the eastern Himalayas, in Bhutan and the Mishmi Hills of Assam where Hodgson's skins originated, and also China.

The closely related Himalayan Tahr or Wild Goat *Hemitragus jemlahicus* was 'found amongst the most inaccessible bare crags of the Hemachal close to the perpetual snows . . . a dauntless and skilful climber . . . a saucy, capricious animal whose freaks of humour and of agility are equally surprising, yet tractable and intelligent . . . soon after his capture (if he be taken young), he becomes content and cheerful; and within a year he may be safely let out, to graze and herd with tame sheep and goats.'[30]

The Bharal or Blue Sheep *Pseudois nayaur* originally described by Hodgson, is a typically Tibetan animal, although it also occurs in the northern regions of Nepal, further north than the Himalayan Tahr. It is regarded as intermediate between a sheep and a goat, although Hodgson considered it a sheep as it was 'a staid, simple, helpless thing which never dreams of transgressing the sobriety of a sheep's nature'.[30] Hodgson was interested in the differences between sheep and goats, and described both physical and behavioural characters of both in his papers.[30,35] However, he believed the only reliable way of separating the two was through a study of their behaviour.

The Argali *Ovis ammon* (Plate 45) is the size of a mule and the largest of all wild sheep. It inhabits the Tibetan plateau, and is reported from trans-Himalayan Nepal. Hodgson wrote: 'They are far more hardy, active and independent than any tame breeds of their kind. They are gregarious . . . leap and run with deer-like power, though as climbers inferior to the Hemitrages (Himalayan Tahr), and as leapers to the Musks. They are often snowed up for days without perishing.'[101]

Perhaps their preference for lower altitudes made the majority of Nepalese deer easier to trap, since Hodgson acquired all six species recorded in Nepal. The Swamp Deer or Barasingha *Cervus duvauceli* which inhabits the edges of large forests and grassy or swampy glades, was actually discovered by Hodgson,[91] although first described by Cuvier possibly from one of the former's specimens. The only high altitude deer in Nepal, the Himalayan Musk Deer *Moschus chrysogaster* (Plate 46) fascinated him. The male has an unusual musk gland which is thought to attract females and is highly valued commercially, causing its serious persecution.[26,81]

Equally as complete as Hodgson's deer collection was that of his lagomorphs (the rabbit family). He found that the Indian Hare *Lepus nigricollis* was 'exceedingly abundant in the Nepalese terai, but less so in the mountains. Hares love the lower and more level tracts within the mountains, where grassy spots are interspersed with copsewood under which they may safely rest and breed. In the plains, patches of grass interspersed with cultivation are the favourite resorts of this species.'[79] A 'fine living pair' of Hispid Hares *Caprolagus hispidus* (Plate 47) was brought to Hodgson from Sikkim. These rare and little known hares inhabit the grass jungles of the terai, and duars of the Himalayan foothills and were seldom seen even in Hodgson's time.[100] He acquired the Royle's Pika *Ochotona roylei*, a gregarious, small lagomorph occurring in the northern region of the Himalayas and Tibet. His specimens came from clefts in rocks on the margin of Gosainkund lake north of Kathmandu.[82]

One of the most extraordinary animals which Hodgson discovered was the Pygmy Hog, the smallest of the pig family: 'about the size of a large Hare and resembling both in form and size a young pig of the ordinary wild kind of about a month old.' It is restricted to the grass jungles at the base of the Himalayas, and is hardly ever seen as it is nocturnal and shy. 'The herds are not large, consisting of 5 or 6, to 15 or 20, and the grown males . . . constantly remain with and defend the females and young. When the annual clearance of the undergrowth of the forest by fire occasionally reveals the Pygmy Hogs, and the herd is thus assailed at advantage, the males with the help of rough and unopen ground really do resist with wonderful energy and frequent success, charging and cutting the naked legs of their human attackers.'[98] He obtained two young ones about six months old 'taken from the nest, a perforation in the bole of a lofty, decayed tree'.

In contrast to the above groups, the highly active and small insectivores must have posed a difficult problem for Hodgson's trappers. Nevertheless, they still collected four species in Nepal and two more in Tibet. Hodgson provided the first description of the Eastern Mole *Talpa micrura*, and noted it was 'very abundant in the Himalaya, the deep bed of black vegetable mould . . . affording a plentiful supply of earthworms which constitute its chief food.'[118]

Considering their fast flight and nocturnal behaviour the bats must have proved equally elusive. It is remarkable that Hodgson managed to obtain about half the number of bat species now known from Nepal. He found about 13 species in the country, mainly in the Kathmandu Valley, as well as another six from India. The Himalayan Leaf-nosed Bat *Hipposideros armiger* and Hodgson's Bat *Myotis formosus* were originally described by him. His papers include a portrayal of the beautiful Greater False Vampire *Megaderma lyra*, to which he gave the much more attractive name of the Slaty-blue Megaderme. He wrote: 'it is extremely gregarious and dwells in the dark parts of houses and outhouses . . . no other species dwells mixedly with this Megaderme . . . entirely nocturnal and insectivorous . . . does not hibernate.'[105]

It is also surprising that Hodgson found 24 of the rodents, two-thirds of the total now recorded in Nepal. His first descriptions of no less than eight of them included the splendid Hodgson's Flying Squirrel *Petaurista magnificus*,[92] (Plate 48) and two other squirrel species.[53] There were also four true rats and mice, as well as the inaptly named Lesser Bamboo Rat *Cannomys badius*. He kept the latter species for several weeks and wrote: 'I never saw such another confident, saucy and yet entirely innocuous creature except it be for the marmot . . . the species lives in small groups in burrows which are usually constructed under the roots of trees . . . roots seemed to be searched for perpetually and they constitute . . . the chief sustenance of the genus.'[85] The Himalayan Crestless Porcupine *Hystrix hodgsoni* was observed as being 'very numerous and very mischievous in the sub-Himalayas where they depredate greatly among . . . edible rooted crops . . . breed in spring and usually produce two young about the time when the crops begin to ripen.'[103]

In addition to his work on Nepalese mammals, Hodgson made an important contribution to the knowledge of Tibetan mammals, collecting 33 species and writing thirteen papers, the first published in 1842.[21,27,80,86,89,95–97,102,106,110,111,113,114,117] Very little was known of the zoology of Tibet at that time, and he provided first descriptions of the Woolly

Hare *Lepus oiostolus*, the Tibetan Fox *Vulpes ferrilata*, the Bharal and the Tibetan Gazelle *Procapra picticauda*. He gave interesting accounts of several species. The first animal Hodgson discovered and wrote about was the Chiru or Tibetan Antelope *Pantholops hodgsoni* in 1826[21] (Plate 49). It occurs mainly in the cold deserts of northern Tibet, but has also recently been reported from trans-Himalayan Nepal. He noted it 'is highly gregarious being usually found in herds of several scores and even hundreds. It is extremely wild and unapproachable by man, to avoid whom it relies chiefly on its wariness and speed. It is very pugnacious and jealous and in its contests often breaks off one of its very long horns. Hence the rumour of Unicorns in Tibet. The Chiru is extremely addicted to the use of salt in the summer months when vast herds are often seen at some of the rock salt beds which so much abound in Tibet.'[109] The Tibetan Gazelle, Hodgson observed was 'An exceedingly graceful little animal . . . found in the deep ravines or low bare hills throughout the plains of middle and eastern Tibet . . . the Goa dwells, either solitarily or in pairs, or at most small families . . . it browses rather than grazes preferring aromatic shrubs to grass.'[95] Other very common inhabitants of the Tibetan plain or plateau included the Kiang, a race of the Asiatic Wild Ass *Equus asinus kiang*, now also recently reported in Nepal, of which Hodgson wrote: 'This exceedingly wild, shy, fleet and handsome species (is found) in herds of moderate size, composed of females and juniors, with seldom above one mature male, and oftener none, except in the breeding season . . . the Tibetans are wholly unable to take it alive, though it is in high esteem amongst them for its beauty and fleetness.'[97] Similarly, the Bobak Marmot was *Marmota bobak* 'very common in the sandy plains of Tibet', although 'rare in the Kachar'. Hodgson kept a few in his garden and noted they were 'very somnolent by day, more active towards night . . . very tame and gentle for the most part . . . live in burrows . . . gregarious . . . each sleeps rolled into a ball and buried in straw . . . hibernates for four months.'[80,89] Mouse-hares or Pikas were so abundant in parts of the country in Hodgson's time that their burrows rendered roads unsafe for horsemen. He collected both the Royle's Pika and also the Black-lipped Pika *Ochotona curzoniae* which he first described.[117]

Hodgson also found several Palearctic mammals in Tibet including the Shou, a race of Red Deer *Cervus elaphus wallichi*, which was 'said to be very generally spread over the wide extent of Tibet . . . it must be considered a Tibetan species only, and not a Himalayan also. Open plains it avoids, frequenting districts more or less mountainous and provided with cover of trees. It is shy and avoids the neighbourhood of villages or houses, but depredates by night upon the outlying crops of barley and wheat.'[114] Other Palearctic species were the Red Fox *Vulpes vulpes*, Eurasian Badger *Meles meles*, Polecat *Mustela putorius*, Stoat *M. erminea*, Lynx *Felis lynx*, and the Grey Wolf *Canis lupus* which he noted as 'common all over Tibet and a terrible depredator of flocks.'[97]

7

A forgotten hero

ALTHOUGH Brian Hodgson was widely acknowledged during his own century as an authority on Buddhism, natural history, and anthropology, there is no doubt his name and achievements today have largely slipped into obscurity. Unless one were a student of nineteenth-century Nepalese history, or had researched into the early developments of those disciplines in which he had played a leading role, his name would be completely unfamiliar. Even when it appears regularly, as in the nomenclature of bird species like Hodgson's Redstart or Hodgson's Bushchat, ornithologists will probably use the name without ever stopping to reflect on the man behind the patronymic. This lack of reputation on Hodgson's part, while it might seem unjustified, can be traced to the influence of several factors.

The century and a half that separates contemporary students of ornithology, anthropology, and Buddhism from the period of Hodgson's own work have meant that much of it is no longer essential to an understanding of any of these disciplines. Rather ironically, in Mahayana Buddhism, of whose study he was acknowledged as the founding father, this is particularly the case. His early work has been buried under a mass of later, more accurate material, and the debt owed to his pioneering efforts has largely been forgotten. Important subsequent changes in the focus of anthropology, meanwhile, have rendered much of his work in this quarter out of date or unfashionable. It is only where other circumstances have conspired to arrest the further development of a subject that contemporary workers have recognized the full importance of Hodgson's achievements. Such a situation existed in the study of Nepalese natural history, where the government's tight restrictions, right up until 1953, on the number of visiting Europeans has allowed only relatively recent advances. It is probably in this field that Hodgson's presence is now most strongly felt.

Even within Hodgson's own life the lapse of time between the period of his greatest efforts and the date of his death had led his obituarists to forget exactly what he had done. One accused him of being 'the greatest authority on Himalayan flora'! This was a subject about which he possibly knew something, but he never published any material on flora, and he was certainly never the greatest authority.[2] Then in *Ibis*, the journal of the British

Ornithologists Union, a short piece squeezed in amongst a host of other announcements, acknowledged only Hodgson's achievements as a collector. A single sentence was deemed sufficient to cover all his papers, and although the existence of the paintings was noted, no mention was made of the extent or excellence of the collection.[1]

Perhaps equally important as the lapse of time as a reason for Hodgson's neglect by others, is the style of his writing. He was educated at a time when classical literature was still very much the measure of all literature, and his writings have the winding and parenthetic character which translations of these dead languages tend to have. A recent editor of Hodgson's political papers described his letters as some of the most turgid prose he had ever had the misfortune to struggle through.[144] Take for example, this single sentence on Himalayan geography:

It is consistent with all we know of the action of these hypogene forces which raise mountains, to suppose that the points of greatest intensity in the pristine action of such forces as marked by the loftiest peaks, should not be surrounded by a proportionate circumjacent intumescence of the general mass; and if there be such an intumescence of the general surface around each preeminent Himalayan peak; it will follow, as clearly in logical sequence as in plain fact it is apparent, that these grand peak-crowned ridges will determine the essential character of the aqueous distribution of the very extended mountain chain (1800 miles) along which they occur at certain palpable and tolerably regular intervals.[120]

No doubt if the language could be broken down, a logical system of thought would be found. Yet its complexity of arrangement is a barrier to the contemporary reader, whose training is for a very different idiom. How much easier to go to a more recent and perhaps derivative source than battle through the original.

In his work on *The Sanskrit Buddhist Literature of Nepal*, Rajendra Lal Mitra said that man formed the centre of Hodgson's studies. The vast amount of work which he did on natural history, however, clearly demonstrates that Hodgson's interests extended beyond this definition. In a sense it is the sheer range of his abilities that has contributed to his obscurity. He divided his attention fairly evenly over a number of disciplines, and made important but not revolutionary contributions to each. Had he devoted himself exclusively to one subject there is no doubt he would have made a larger impact in that field. He was never of the stature of naturalists like Darwin or Huxley, nor of an Orientalist like Müller. His writings, largely in essay form, were regular, significant, and accurate building blocks which added to and modified what was previously known, but he produced no single definitive or radical publication such as Darwin's *On the Origin of Species*.

Implicit in so much of Hodgson's work was the belief that learning and knowledge were best gained by the action of many workers, for the benefit, not just of individuals, but for society as a whole. He clearly intended his essays on Buddhism, for example, to be seminal, and hoped that they would spark off in others a number of fruitful responses. It is in his collecting, however, that his assumption about the social nature of the search for knowledge is most apparent. The manuscripts and natural historical specimens, which he amassed at considerable expense and with indefatigable energy, once gathered together, were distributed widely and freely with no less enthusiasm for use by others. It is interesting to compare Hodgson's own open-handed donations of his skins with the calculating methods of John Gould. The latter, piqued by the British Museum's refusal to pay the £1000 he asked for his Australasian specimens, sold them to the Academy of Natural Sciences of Philadelphia.[142]

Hodgson was no less free with his discoveries than he was with his collections. When the same Gould offered such self-centred terms for his participation in a joint work on Nepalese zoology, although Hodgson refused to comply, he was still willing, if no other means of publication presented itself, to hand the materials over to the artist for his private use. Why should Hodgson offer the fruits of his research to one who had so clearly sought to make financial and personal gain from their partnership, unless he had hoped that the ends of science were somehow served by so doing?

In any discussion of reputation there is a tacit assumption that public acclaim is both a valuable personal reward and an important measurement of achievement and ability. However, there is no definite evidence that Hodgson ever agreed with these assumptions, or that he was ever eager to be famous. He showed a natural human anxiety where he thought credit should be given, and felt let down when others seemed unwilling to acknowledge what they owed to him. This is clearly displayed in Hodgson's Buddhist essays, and in his later natural history papers. Yet his reason for writing these papers never seems to have been solely for personal ends. In fact, there are examples where this is obviously not the case. In 1858 when he was forced by family considerations to return to England and abandon his old ambition to write and publish a work on Nepalese history, he did so without regret. Testimony to his generous acceptance of such a fate are the 90 volumes of manuscripts and personal writings which he gave to the India Office Library, and which are presently housed with the British Library in London. Any person whose concern was purely for fame would have worked up the substantial materials he already possessed; yet Hodgson preferred to leave them for the benefit of others, rather than prepare an incomplete or inaccurate work.

In spite of the number of reasons advanced here to explain why Hodgson was never widely recognized, and setting aside his own considerations on the matter, anyone who becomes acquainted with the qualities and achievements of a person like Brian Hodgson cannot escape the conviction that he was unjustly overlooked. The *Ibis* obituary said that 'every mark of distinction which the learned societies of Europe could confer was deservedly bestowed upon Hodgson but as might be expected he was never knighted nor asked to become a member of the House of Lords'.[1] When one considers all his services, not only to the British, but also to the Indian and Nepalese peoples, it is impossible to understand why he did not receive both honours. Men whose accomplishments were half those of Hodgson had received more. It was not only the British government who neglected him either. He had to reach his ninetieth year before Oxford University deigned to confer an honorary doctorate upon him. A bust, probably a copy of that prepared by the Asiatic Society of Bengal, and presently sitting in a dusty cupboard in the British Museum, is an appropriate image of his neglect by workers in the natural sciences. It is almost as if Hodgson was destined to be forgotten. In a very recent publication on the history of the Gurkha troops, Hodgson's very early advocacy of their employment, and his crucial part in the events of 1857, when employment of the Nepalese army to help quell the mutiny first opened the doors to regular recruitment, are both completely ignored.[12]

However, it is inappropriate at the end of this book to close on a note of resentment or frustration on Hodgson's behalf, for that is not how he himself would have seen things. The most fitting image is that of an old man, in his nineties, in the shade of his Gloucestershire garden, patiently and carefully revising his essays: the picture of a man's search for truth, indifferent alike to the praise or blame of the outside world.

APPENDIX

Bird species discovered by Hodgson

Here * indicates bird species first described for science by Hodgson. Other bird species listed were discovered by Hodgson but their first descriptions are attributed to others.

* Mountain Hawk-Eagle *Spizaetus nipalensis*
* Snow Partridge *Lerwa lerwa*
* Tibetan Partridge *Perdix hodgsoniae*
* White Eared Pheasant *Crossoptilon crossoptilon*
  Ibisbill *Ibidorhyncha struthersii*
  Long-billed Plover *Charadrius placidus*
* Solitary Snipe *Gallinago solitaria*
* Wood Snipe *Gallinago nemoricola*
  Speckled Wood pigeon *Columba hodgsonii*
  Oriental Cuckoo *Cuculus saturatus*
* Oriental Scops Owl *Otus sunia*
* Forest Eagle Owl *Bubo nipalensis*
* Tawny Fish Owl *Ketupa flavipes*
  Hodgson's Frogmouth *Batrachostomus hodgsoni*
  Dark-rumped Swift *Apus acuticauda*
* Rufous-necked Hornbill *Aceros nipalensis*
* White-browed Piculet *Sasia ochracea*
* Bay Woodpecker *Blythipicus pyrrhotis*
  Crimson-breasted Woodpecker *Dendrocopos cathpharius*
  Silver-breasted Broadbill *Serilophus lunatus*
* Blue-naped Pitta *Pitta nipalensis*
  Nepal House Martin *Delichon nipalensis*
* Upland Pipit *Anthus sylvanus*
  Rosy Pipit *Anthus roseatus*
* Black-winged Cuckoo-shrike *Coracina melaschistos*
  Ashy Bulbul *Hypsipetes flavalus*
* Maroon-backed Accentor *Prunella immaculata*

Rufous-breasted Accentor *Prunella strophiata*
Robin Accentor *Prunella rubeculoides*
* Indian Blue Robin *Luscinia brunnea*
* Golden Bush Robin *Tarsiger chrysaeus*
  Hodgson's Redstart *Phoenicurus hodgsoni*
  White-throated Redstart *Phoenicurus schisticeps*
  White-bellied Redstart *Hodgsonius phoenicuroides*
* White-tailed Robin *Cinclidium leucurum*
* Grandala *Grandala coelicolor*
* Purple Cochoa *Cochoa purpurea*
* Green Cochoa *Cochoa viridis*
  Hodgson's Bushchat *Saxicola insignis*
  Grey Bushchat *Saxicola ferrea*
  Chestnut Thrush *Turdus rubrocanus*
* Black-backed Forktail *Enicurus immaculatus*
* Slaty-backed Forktail *Enicurus schistaceus*
* Grey-bellied Tesia *Tesia cyaniventer*
* Brown-flanked Bush Warbler *Cettia fortipes*
  Chestnut-crowned Bush Warbler *Cettia major*
* Aberrant Bush Warbler *Cettia flavolivacea*
* Grey-sided Bush Warbler *Cettia brunnifrons*
  Spotted Bush Warbler *Bradypterus thoracicus*
* Brown Bush Warbler *Bradypterus luteoventris*
* Striated Prinia *Prinia criniger*
  Grey-capped Prinia *Prinia cinereocapilla*

White-spectacled Warbler *Seicercus affinis*

* Chestnut-crowned Warbler *Seicercus castaniceps*

Grey-hooded Warbler *Seicercus xanthoschistos*

Broad-billed Warbler *Abroscopus hodgsoni*

Rufous-faced Warbler *Abroscopus albogularis*

Black-faced Warbler *Abroscopus schisticeps*

Orange-barred Leaf Warbler *Phylloscopus pulcher*

Grey-faced Leaf Warbler *Phylloscopus maculipennis*

* Smoky Warbler *Phylloscopus fuligiventer*

* Rufous-bellied Niltava *Niltava sundara*

Pygmy Blue Flycatcher *Muscicapella hodgsoni*

* Ferruginous Flycatcher *Muscicapa ferruginea*

* Slaty-blue Flycatcher *Ficedula tricolor*

* White-gorgetted Flycatcher *Ficedula monileger*

* Orange-gorgetted Flycatcher *Ficedula strophiata*

* White-browed Scimitar-Babbler *Pomatorhinus schisticeps*

* Streak-breasted Scimitar-Babbler *Pomatorhinus ruficollis*

* Greater Scaly-breasted Wren-Babbler *Pnoepyga albiventer*

* Lesser Scaly-breasted Wren-Babbler *Pnoepyga pusilla*

Black-chinned Babbler *Stachyris pyrrhops*

Golden Babbler *Stachyris chrysaea*

Grey-throated Babbler *Stachyris nigriceps*

* Great Parrotbill *Conostoma aemodium*

* Brown Parrotbill *Paradoxornis unicolor*

Black-breasted Parrotbill *Paradoxornis flavirostris*

* Fulvous Parrotbill *Paradoxornis fulvifrons*

* Black-throated Parrotbill *Paradoxornis nipalensis*

Spiny Babbler *Turdoides nipalensis*

Slender-billed Babbler *Turdoides longirostris*

* White-throated Laughing-thrush *Garrulax albogularis*

* Lesser Necklaced Laughing-thrush *Garrulax monileger*

* Greater Necklaced Laughing-thrush *Garrulax pectoralis*

* Grey-sided Laughing-thrush *Garrulax caerulatus*

Scaly Laughing-thrush *Garrulax subunicolor*

Black-faced Laughing-thrush *Garrulax affinis*

* Silver-eared Mesia *Leiothrix argentauris*

Fire-tailed Myzornis *Myzornis pyrrhoura*

* Cutia *Cutia nipalensis*

Green Shrike-Babbler *Pteruthius xanthochloris*

* Black-eared Shrike-Babbler *Pteruthius melanotis*

* Hoary Barwing *Actinodura nipalensis*

* Blue-winged Minla *Minla cyanouroptera*

* Chestnut-tailed Minla *Minla strigula*

* Red-tailed Minla *Minla ignotincta*

Golden-breasted Fulvetta *Alcippe chrysotis*

* Rufous-winged Fulvetta *Alcippe castaneceps*

* White-browed Fulvetta *Alcippe vinipectus*

* Nepal Fulvetta *Alcippe nipalensis*

* Long-tailed Sibia *Heterophasia picaoides*

* Whiskered Yuhina *Yuhina flavicollis*

* Stripe-throated Yuhina *Yuhina gularis*

* Rufous-vented Yuhina *Yuhina occipitalis*

* Black-chinned Yuhina *Yuhina nigrimenta*

* White-bellied Yuhina *Yuhina zantholeuca*

Black-browed Tit *Aegithalos iouschistos*

Grey-crested Tit *Parus dichrous*

* Sultan Tit *Melanochlora sultanea*

Rusty-flanked Treecreeper *Certhia nipalensis*

* Green-tailed Sunbird *Aethopyga nipalensis*

* Black-throated Sunbird *Aethopyga saturata*

* Fire-tailed Sunbird *Aethopyga ignicauda*

* Streaked Spiderhunter *Arachnothera magna*

Yellow-bellied Flowerpecker *Dicaeum melanoxanthum*

Buff-bellied Flowerpecker *Dicaeum ignipectus*

* Crow-billed Drongo *Dicrurus annectans*
* Plain Mountain-Finch *Leucosticte nemoricola*
* Dark-breasted Rosefinch *Carpodacus nipalensis*

Beautiful Rosefinch *Carpodacus pulcherrimus*

Red-breasted Rosefinch *Carpodacus puniceus*

* Crimson-browed Finch *Propyrrhula subhimachala*
* Scarlet Finch *Haematospiza sipahi*
* Gold-naped Finch *Pyrrhoplectes epauletta*
* Brown Bullfinch *Pyrrhula nipalensis*
* Spot-winged Grosbeak *Mycerobas melanozanthos*
* White-winged Grosbeak *Mycerobas carnipes*

BIBLIOGRAPHY

1. Anon. Obituary of B. H. Hodgson. *Ibis* **6**(6), 580–1 (1894).
2. Anon. Biographical Note. In *Encyclopaedia Britannica*, 11th edn, Vol. 13, p. 557. Encyclopaedia Britannica Company, New York (1910).
3. Anon. *National conservation strategy for Nepal: a prospectus.* International Union for Conservation of Nature and Natural Resources and His Majesty's Government of Nepal. Gland, Switzerland (1983).
4. Bailey, F. M. Register of bird specimens collected in Nepal 1935–38, and presented to the British Museum (Natural History). Unpublished (1938).
5. Biswas, B. The birds of Nepal. *J. Bombay nat. Hist. Soc.* **57**(2), 278–308 (1960).
6. —— The birds of Nepal, Part 3. *J. Bombay nat. Hist. Soc.* **58**(2), 444–74 (1961).
7. Blyth, E. List of birds obtained in the vicinity of Calcutta, from September 1841 to March 1843 inclusive. *Ann. Mag. nat. Hist.* **12**, 90–101 (1844).
8. —— Appendix to the report for December meeting, 1842. *J. Asiat. Soc. Bengal* **13**, 361–95 (1844).
9. Cantor, T. E. Spicilegium Serpenticum Indicorum. *Proc. zool. Soc. London.* 31–4, 49–55 (1839).
10. Corbet, G. B. and Hill, J. E. *A world list of mammalian species.* British Museum (Natural History), London (1978).
11. Desmond, R. *The Indian Museum 1801–1879.* HMSO, London (1982).
12. Farwell, B. *Gurkhas.* Penguin Books, Harmondsworth (1985).
13. Fleming, R. L. Sr. and Traylor, M. A. Notes on Nepal birds. *Fieldiana: zool.* **35**(9), 447–87 (1961).
14. Fürer-Haimendorf, C. von. (Ed). *Contributions to the anthropology of Nepal: Proceedings of a symposium held at the School of Oriental and African Studies*, University of London in 1973. Aris & Phillips, Warminster (1974).
15. Gould, J. *A century of birds from the Himalayan mountains.* J. Gould, London (1834).
16. Gould, J. *Letter to Brian Hodgson (Senior).* 6 March 1837. Stored in the Zoology Library, British Museum (Natural History), London.
17. Gray, J. E. *Catalogue of the specimens and drawings of mammals, birds, reptiles and fishes of Nepal and Tibet, presented by B. H. Hodgson, Esq. to the British Museum*, 2nd edn. London (1863).
18. Gray, J. E. and Gray, G. R. *Catalogue of the specimens and drawings of mammalia and birds of Nepal and Thibet, presented by B. H. Hodgson, Esq. to the British Museum.* London (1846).
19. Günther, A. Contributions to a knowledge of the reptiles of the Himalaya Mountains. *Proc. zool. Soc. London* 148–75 (1860).

20. Günther, A. List of the cold-blooded Vertebrata collected by B. H. Hodgson Esq., in Nepal. *Proc. zool. Soc. London* 213–27 (1861).

21. Hodgson, B. H. Account of the Chiru or Unicorn of the Himalaya mountains (Pantholops Hodgsoni). Tilloch, *Phil. Mag.* **68**, 232–4 (1826).

22. —— On a new species of Buceros. *Gleanings in Science* **1**, 249–52 (1829).

23. —— On the Chiru or Antilope Hodgsonii, Abel. *Gleanings in Science* **2**, 348–52 (1830).

24. —— On the Bubaline Antelope (Antilope Thar). *Gleanings in Science* **3**, 122–3 (1831).

25. —— On some of the Scolopacidae of Nepal. *Gleanings in Science* **3**, 233–43 (1831).

26. —— Contributions in natural history (the Musk Deer and Cervus Jarai). *Gleanings in Science* **3**, 320–4 (1831).

27. —— Description and characters of the Chiru Antelope (Antilope Hodgsonii, Abel). *Proc. zool. Soc. London* **1**, 52–4 (1831).

28. —— On a species of Aquila, Circaeetus and Dicrurus. *Asiat. Res.* **18**(2), 13–26 (1833).

29. —— On the migration of the Natatores and Grallatores, as observed at Kathmandu. *Asiat. Res.* **18**(2), 122–128 (1833).

30. —— The Wild Goat (Capra Jharal) and the Wild Sheep of Nepal (Ovis Nayaur). *Asiat. Res.* **18**(2), 129–38 (1833).

31. —— Description of the Buceros Homrai of the Himalaya. *Asiat. Res.* **18**(2), 169–88 (1833).

32. —— Description of the Wild Dog of the Himalaya (Canis primaevus). *Asiat. Res.* **18**(2), 221–37 (1833).

33. —— Characters of a new Species of Perdix. *Proc. zool. Soc. London* **1**, 107 (1833).

34. —— Letter on the distinction between the Ghoral (Antilope Goral, Hardw.) and Thar (Antilope Thar, Hodgs.). *Proc. zool. Soc. London* **2**, 85–7 (1834).

35. —— On the characters of the Jharal (Capra Jharal, Hodgs.), and of the Nahoor (Ovis Nahoor, Hodgs.), with observations on the distinction between the genera Capra and Ovis. *Proc. zool. Soc. London* **2**, 106–9 (1834).

36. —— Letter to James Prinsep, April 1835. Stored in the India Office Library, London.

37. —— Letter to Sir Alexander Johnstone, 20 June 1835. Stored in the Zoology Library, British Museum (Natural History), London.

38. —— Red-billed Erolia. *J. Asiat. Soc. Bengal* **4**, 458–61 (1835).

39. —— Note on the Red-billed Erolia. *J. Asiat. Soc. Bengal* **4**, 701–2 (1835).

40. —— Letter to James Prinsep, 26 March 1836. Stored in the India Office Library, London (1836).

41. —— Description of three new species of Paradoxurus (P. hirsutus, P. Nepalensis, P. lanigerus). *Asiat. Res.* **19**, 72–86 (1836).

42. —— Notices of the ornithology of Nepal. 1. Eight New Species of Cinclosoma. *Asiat. Res.* **19**, 143–50 (1836).

43. Hodgson, B. H. Notices of the ornithology of Nepal. 2. New Species of the Thick billed Finches. *Asiat. Res.* **19**, 150–9 (1836).

44. —— Notices of the ornithology of Nepal. 3. New Genera of the Columbidae. *Asiat. Res.* **19**, 159–64 (1836).

45. —— Notices of the ornithology of Nepal. 4. New Genus and 3 New Species of the Silviadae. *Asiat. Res.* **19**, 165–7 (1837). *J. Asiat. Soc. Bengal* **6**, 230–2 (1836).

46. —— Notices of the ornithology of Nepal. 5. New Species of the Strigine family. *Asiat. Res.* **19**, 168–77 (1836).

47. —— Notices of the ornithology of Nepal. 7. Two New Species of the Parrot Tribe. *Asiat. Res.* **19**, 177–8 (1836).

48. —— Notices of the ornithology of Nepal. 8. New Species of Pomatorhinus, and its Allies. *Asiat. Res.* **19**, 179–186 (1836).

49. —— Notices of the Ornithology of Nepal. 9. New Species of Motacillinae. *Asiat. Res.* **19**, 186–92 (1836).

50. —— Summary Description of some New Species of Birds of Prey. *Bengal Sporting Mag.* **8**, 177–83 (1836).

51. —— - Description of a New Species of Columba. *J. Asiat. Soc. Bengal* **5**, 122–4 (1836).

52. —— Summary Description of some New Species of Falconidae. *J. Asiat. Soc. Bengal* **5**, 227–31 (1836).

53. —— Synoptical description of sundry new animals enumerated in the catalogue of Nepalese mammals. *J. Asiat. Soc. Bengal* **5**, 231–8 (1836).

54. —— Description of two new Species belonging to a new form of the Meruline Group of Birds, with indication of their generic character. *J. Asiat. Soc. Bengal* **5**, 358–60 (1836).

55. —— On a new Genus of the Meropidae. *J. Asiat. Soc. Bengal* **5**, 360–62 (1836).

56. —— On a new Piscatory Genus of the Strigine Family. *J. Asiat. Soc. Bengal* **5**, 363–5 (1836).

57. —— Additions to the ornithology of Nepal. Indication of a new Genus of Insessorial Birds. *J. Asiat. Soc. Bengal* **5**, 770–5 (1837); **6**, 110–12 (1836).

58. —— Additions to the ornithology of Nepal. Indication of a new Genus of Waders, belonging to the Charadriatic Family. *J. Asiat. Soc. Bengal* **5**, 775–7 (1836).

59. —— Additions to the Ornithology of Nepal. Indication of a new Genus of the Picidae, with description of the type. A new species, also, of two new species of the Genus Sitta. *J. Asiat. Soc. Bengal* **5**, 778–9 (1836).

60. —— Additions to the Ornithology of Nepal. New species of Hirundinidae. *J. Asiat. Soc. Bengal* **5**, 779–81 (1836).

61. Hodgson, B. (Senior). Letter to John Gould, 10 March 1837. Stored in the Zoology Library, British Museum (Natural History), London (1837).

62. Hodgson, B. H. Indian Quails. *Bengal Sporting Mag.* **9**, 343–6 (1837).

63. Hodgson B. H. On some New Species of the Edolian and Ceblepyrine subfamilies of the Laniidae of Nepal. *India Review* **1**, 325–8 (1837).

64. —— On some New Species of the more typical Laniidae of Nepal. *India Review* **1**, 445–7 (1837).

65. —— Indication of a new Genus of Insessores, tending to connect the Sylviadae and Muscicapidae. *India Review* **1**, 650–2 (1837).

66. —— On three new Genera or sub-Genera of long-legged Thrushes, with descriptions of their species. *J. Asiat. Soc. Bengal* **6**, 101–4 (1837).

67. —— Description of three new species of Woodpecker. *J. Asiat. Soc. Bengal* **6**, 104–9 (1837).

68. —— On some new Genera of Raptores, with remarks on the old genera. *J. Asiat. Soc. Bengal* **6**, 361–73 (1837).

69. —— On the Bibos, Gauri Gau or Gaurika Gau of the Indian forests. (Bibos cavifrons, B. classicus, B. Aristotelis) *J. Asiat. Soc. Bengal* **6**(2), 745–9 (1837).

70. —— Indication of a new Genus belonging to the Strigine Family, with Description of the New Species and Type. *Madras J. Lit. Sci.* **5**, 23–5 (1837).

71. —— On the Structure and Habits of the Elanus melanopterus. *Madras J. Lit. Sci.* **5**, 75–8 (1837).

72. —— On two new Genera of Rasorial Birds. *Madras J. Lit. Sci.* **5**, 300–5 (1837).

73. —— Indication of some new forms belonging to the Parianae. *India Review* **2**, 30–34, 87–90 (1838).

74. —— On a new species of pheasant Phasianus crossoptilon from Tibet. *J. Asiat. Soc. Bengal* **7**, 863–5 (1838).

75. —— On a new Genus of the Fissirostral Tribe. *J. Asiat. Soc. Bengal* **8**, 35–6 (1839).

76. —— Description of two new Species of a new form of Meruline Birds. *J. Asiat. Soc. Bengal* **8**, 37–8 (1839).

77. —— On Cuculus. *J. Asiat. Soc. Bengal* **8**, 136–7 (1839).

78. —— Letter to James Prinsep, 4 October 1840. Stored in the British Library, London (1840).

79. —— On the Common Hare of the Gangetic Provinces, and of the Sub-Himalaya with a slight notice of a strictly Himalayan species. *J. Asiat. Soc. Bengal* **9**, 1183–6 (1840).

80. —— Notice of the Marmot (Arctomys Himalayanus) of the Himalaya and Tibet. *J. Asiat. Soc. Bengal* **10**, 777–8 (1841).

81. —— On a new organ in the genus Moschus. *J. Asiat. Soc. Bengal* **10**, 795–6 (1841).

82. —— On a new species of Lagomys inhabiting Nepal (Lagomys Nepalensis, nobis). *J. Asiat. Soc. Bengal* **10**, 854–5 (1841).

83. —— Notice of a new form of the Glaucopinae, or Rasorial Crows, inhabiting the Northern region of Nepal—Conostoma Aemodius (Nobis type). *J. Asiat. Soc. Bengal* **10**, 856–7 (1841).

84. Hodgson B. H. On the Civet of the continent of India, Viverra orientalis (hodie melanurus). *Calcutta J. nat. Hist.* **2**, 47–56 (1842).

85. —— New species of Rhizomys discovered in Nepal (R. badius Bay Bamboo Rat). *Calcutta J. nat. Hist.* **2**, 60–2, 410–11 (1842).

86. —— Notice of the mammals of Tibet. *J. Asiat. Soc. Bengal* **11**, 275–89 (1842).

87. —— Letter to Mr Hawkins, 12 August 1843. Stored in the Zoology Library, British Museum (Natural History), London.

88. —— Description of a new genus of Falconidae. *J. Asiat. Soc. Bengal* **12**, 127–8 (1843).

89. —— Notice of two marmots found in Tibet (Arctomys Himalayanus of Catalogue, potius Tibetensis, hodie mihi, and A. Hemachalanus). *J. Asiat. Soc. Bengal* **12**, 409–14 (1843).

90. —— Additions to the Catalogue of Nepal Birds. *J. Asiat. Soc. Bengal* **12**, 447–50 (1843).

91. —— On a new species of Cervus (C. dimorphe) *J. Asiat. Soc. Bengal* **12**(2), 897 (1843).

92. —— Summary description of two new species of flying squirrel (Sciuropterus chryostrix, Sc. senex). *J. Asiat. Soc. Bengal* **13**, 67–8 (1844).

93. —— *Catalogue of Nipalese Birds, collected between 1824 and 1844.* In Gray, J. E., *Zoological Miscellany*, June (1844).

94. —— On Nepalese birds *Proc. zool. Soc. London* **13**, 22–37 (1845).

95. —— Description of a new species of Tibetan Antelope (Procapra picticaudata). *J. Asiat. Soc. Bengal* **15**, 334–8 (1846).

96. —— On the wild sheep of Tibet, with plates (Ovis Ammonoides, mihi). *J. Asiat. Soc. Bengal.* **15**, 338–43 (1846).

97. —— Description of the Wild Ass (Asinus Polyodon) and Wolf of Tibet (Lupus Laniger) *Calcutta J. nat. Hist.* **7**, 469–77 (1847).

98. —— On a new form of the Hog Kind or Suidae (Porcula Salvania, Pigmy Hog). *J. Asiat. Soc. Bengal* **16**, 423–8 (1847).

99. —— Description of *Pteruthius melanotis* In Blyth. *J. Asiat. Soc. Bengal* **16**, 448 (1847).

100. —— On the Hispid Hare of the Saul forest (Lepus hispidus, Pears., Caprolagus hispidus, Blyth). *J. Asiat. Soc. Bengal* **16**, 572–7 (1847).

101. —— On various genera of the ruminants. *J. Asiat. Soc. Bengal* **16**, 685–711 (1847).

102. —— On the Tibetan Badger (Taxidia leucurus). *J. Asiat. Soc. Bengal* **16**, 763–71 (1847).

103. —— On a new species of Porcupine (Hystrix alophus). *J. Asiat. Soc. Bengal* **16**, 771–4 (1847).

104. —— On the Charj or Otis Bengalensis. *J. Asiat. Soc. Bengal* **16**, 883–9.

105. —— The Slaty-blue Megaderme (M. schistacea). *J. Asiat. Soc. Bengal* **16**, 889–94 (1847).

106. —— On the tame sheep and goats of the Sub-Himalayas and Tibet. *J. Asiat. Soc. Bengal* **16**, 1003–26 (1847).

107. Hodgson B. H. On the cat-toed Subplantigrades of the Sub-Himalayas. *J. Asiat. Soc. Bengal* **16**, 1113–29 (1847).

108. —— Observations on the manners and structure of Prionodon pardicolor *Calcutta J. nat. Hist.* **8**, 40–5 (1848).

109. —— On the four-horned antelopes of India. *Calcutta J. nat. Hist.* **8**, 87–94 (1848).

110. —— Note on the Kiang. *Calcutta J. nat. Hist.* **8**, 98–100 (1848).

111. —— The Polecat of Tibet, n.s. *J. Asiat. Soc. Bengal* **18**, 446–50 (1849).

112. —— On the Takin (Budorcas taxicolor) of the Eastern Himalaya. *J. Asiat. Soc. Bengal* **19**, 65–75 (1850).

113. —— On the Shou or Tibetan Stag (Cervus affinis). *J. Asiat. Soc. Bengal* **19**, 466–9, 518–20 (1850).

114. —— On the Shou or Tibetan Stag (C. Affinis mihi). *J. Asiat. Soc. Bengal* **20**, 388–94 (1851).

115. —— On the Geographical Distribution of the Mammalia and Birds of the Himalaya. *Proc. zool. Soc. London* **23**, 124–8 (1855).

116. —— On a new Perdicine bird (Sacpha Hodgsoniae) from Tibet. *J. Asiat. Soc. Bengal* **25**, 165–6 (1856).

117. —— On a new species of Lagomys (L. Curzoniae) and a new Mustela (M. Temnon) inhabiting the north of Sikhim and the proximate parts of Tibet. *J. Asiat. Soc. Bengal* **26**, 207–8 (1857).

118. —— Description of a new species of Himalayan Mole (Talpa Macrura). *J. Asiat. Soc. Bengal* **27**, 176 (1858).

119. —— Letter to Mr Hawkins, 10 December 1859. Stored in the Zoology Library, British Museum (Natural History), London.

120. —— *Essays on the languages, literature, and religion of Nepal and Tibet: together with further papers on the geography, ethnology, and the commerce of those countries.* Trubner, London (1874).

121. —— *Miscellaneous essays relating to Indian subjects*, in two volumes. Trubner, London (1880).

122. —— Notes and original watercolour paintings of the birds of Nepal, Tibet and India, held in the Zoological Society of London Library. Unpublished. Undated.

123. —— Notes and original water colour paintings of the mammals of Nepal, Tibet and India, held in the Zoological Society of London Library. Unpublished. Undated.

124. —— Private note on John Gould's letter of 6 March 1837. Undated. Stored in the Zoology Library, British Museum (Natural History), London (1837).

125. Horsfield, T. and Moore, F. *A catalogue of birds in the Museum of the Hon. East-India Company.* W. H. Allen, London (1854).

126. Hume, A. O. and Oates, E. W. *The nests and eggs of Indian birds*, 3 volumes, 2nd edn. Porter, London (1890).

127. Hunter, W. W. *Life of Brian Houghton Hodgson.* John Murray, London (1896).

128. Huxley, L. *The life and letters of Sir Joseph Dalton Hooker*, Vol I. John Murray, London (1918).

129. Inskipp, C. and Inskipp, T. *A guide to the birds of Nepal.* Croom Helm, London (1985).

130. Inskipp, T. P. *et al.* Notes on birds recorded in Nepal, September 1970 to March 1971. Unpublished. (1971).

131. Jerdon, T. C. Supplementary notes to 'The Birds of India'. *Ibis* **1**(3), 234–47; 335–6 (1871).

132. Krabbe, E. List of bird specimens in the Zoological Museum of Copenhagen, collected by G. B. Gurung, S. Rana, and P. W. Soman from Nepal, 1959. Unpublished. (1983).

133. Mills, D. G. H. and Preston, N. A. Notes on birds recorded in Nepal 1981. Unpublished. (1981).

134. Mitra, R. L. *The Sanskrit Buddhist literature of Nepal.* J. W. Thomas Baptist Mission Press, Calcutta (1882).

135. Porter, R. F., Oddie, W. E. and Marr, B. A. E. Notes on birds recorded in Nepal February 1981. Unpublished. (1981).

136. Prinsep, J. Letter to Sir Alexander Johnstone, 6 August 1837. Stored in the Zoology Library, British Museum (Natural History), London.

137. Proud, D. Some notes on the birds of the Nepal Valley. *J. Bombay nat. Hist. Soc.* **48**, 695–719 (1949).

138. —— Notes on some Nepalese birds. *J. Bombay nat. Hist. Soc.* **58**, 277–9 (1961).

139. Ripley, S. D. Birds from Nepal 1947–49. *J. Bombay nat. Hist. Soc.* **49**, 355–417 (1950).

140. Sanwal, B. D. *Nepal and the East India Company.* Bombay Chronicle Press, London (1965).

141. Scully, J. A contribution to the ornithology of Nepal. *Stray Feathers* **8**, 204–368 (1879).

142. Sharpe, R. Bowdler. *An analytical index to the works of the late John Gould, F.R.S.; with a biographical memoir and portrait.* Sotheran, London (1893).

143. —— *The history of the collections contained in the natural history departments of the British Museum* II. British Museum (Natural History), London (1906).

144. Stiller, L. *Letters from Kathmandu: the Kot Massacre.* Tribhuvan University Press, Kathmandu (1981).

145. Swan, L. W. and Leviton, A. E. The herpetology of Nepal: a history, checklist and zoogeographical analysis of the herpetofauna. *J. Bengal nat. Hist. Soc.* **34**(2), 91–144 (1966).

146. Torrens, H. Letter to B. H. Hodgson, 21 September 1844. Stored in the Zoological Society of London Library.

147. Warren, R. L. M. and Harrison, C. J. O. *Type-specimens of birds in the British Museum (Natural History),* Vol. 2, *Passerines.* British Museum (Natural History), London (1971).

148. Welbon, G. R. *The Buddhist Nirvana and its Western interpreters.* Chicago University Press, Chicago (1968).

Index